SILLIMAN MEMORIAL LECTURES

Radioactive Transformations

BY

Ernest Rutherford

With a New Foreword by

Frank Wilczek

Yale UNIVERSITY PRESS

NEW HAVEN & LONDON

Yale University Press books may be purchased in quantity for educational, business, or promotional use. For information, please e-mail sales.press@yale.edu (U.S. office) or sales@yaleup.co.uk (U.K. office).

Library of Congress Control Number: 2012937089

ISBN 978-0-300-18130-2 (pbk.)

Printed in the United States of America.

A catalogue record for this book is available from the British Library.

This paper meets the requirements of ANSI/NISO Z39.48-1992 (Permanence of Paper).

10 9 8 7 6 5 4 3 2 1

publication of this book is enabled by a grant from

Figure Foundation

THE SILLIMAN FOUNDATION LECTURES

On the foundation established in memory of Mrs. Hepsa Ely Silliman, the President and Fellows of Yale University present an annual course of lectures designed to illustrate the presence and providence of God as manifested in the natural and moral world. It was the belief of the testator that any orderly presentation of the facts of nature or history contributed to this end more effectively than dogmatic or polemical theology, which should therefore be excluded from the scope of the lectures. The subjects are selected rather from the domains of natural science and history, giving special prominence to astronomy, chemistry, geology, and anatomy.

Contents

Foreword

Ernest Rutherford's *Radioactive Transformations* is, of course, a scientific document, but that is no longer its primary interest. The discoveries it announces have long since been assimilated into textbooks and appear as special cases within much more comprehensive and coherent bodies of knowledge. Nevertheless we can read it with pleasure and profit today, as a remarkable piece of literature combining traveler's tale, historical chronicle, and accidental autobiography.

As a traveler's tale, it is a richly detailed description of a strange new world, a world distilled from ours by weird and laborious procedures the old alchemists could call their own, a world of causeless transformations and bizarre emanations.

As a historical chronicle, it recounts an epoch when "discoveries of the most striking interest and importance have followed one another in rapid succession. . . . The march of discovery has been so rapid that it has been difficult even for those directly engaged in the investigations to grasp at once the full significance of the facts that have been brought to light" (page 1). In retrospect, this era appears as the time when physics first truly came to grips with the issue of what matter *is,* as opposed to how matter, being given, behaves. It was a time of suddenly expanding horizons, awakened ambitions, and triumphal achievement.

As accidental autobiography, *Radioactive Transformations* is a work in which Rutherford speaks not a personal word, yet from its pages a remarkable, and remarkably attractive, personality emerges. He is a real-life Sherlock Holmes, fastening on odd facts, theorizing within their discipline, testing his intuitions relentlessly. Yet in his candor and simplicity of character, and what appears in retrospect as occasional theoretical naiveté, there's also a leavening pinch of Doctor Watson.

Radioactive Transformations is based on Ernest Rutherford's Silliman Lectures for 1905. Following J. J. Thomson's discovery of the electron in 1897 and Henri Becquerel's discovery of radioactivity in 1896, suddenly atoms were no longer the ideal objects posited by ancient philosophy and contemporary theory: indivisible, unchanging, and therefore immune to analysis. No! Atoms had parts, and they changed, spontaneously, in peculiar, seemingly whimsical ways. Lacking ultimate simplicity, atoms could—and for a man like Rutherford, that meant they *must*—be understood more deeply. They must be looked into, and taken apart, until their inner logic was revealed. The theoretical physics of the time was, manifestly, not up to the job. It was a heady time for experimental physics, perfectly suited to Rutherford's ingenuity, energy, and ambition.

The word *revolution* is overused in the history of science. Indeed, a primary virtue of science is that its laws are rooted in evidence, and subject to continuous scrutiny and amendment. Scientific ideas and models of long standing have been tested and strengthened over time. Therefore truly radical changes, as opposed to additions and refinements, are rare. Yet in 1905, in physics, Rutherford was creatively engaged with two genuinely revolutionary developments: intimations of the divisibility, and of the instability, of the basic constituents of matter.

Today atomic and nuclear physics are mature subjects,

supported in dense, seamless webs of facts and ideas. They allow long chains of confident deduction and enable complex engineering projects. In other words, they have been domesticated. But in 1905 they were wild. The first chapter of *Radioactive Transformations,* entitled "Historical Introduction," is Rutherford's take on the state of play. It is a fascinating account, not to be missed, but it is aimed toward a reader who no longer quite exists: someone sophisticated in the physics and experimental technology of 1905, yet innocent of many basic (later) discoveries that children learn today. So a few words of orientation and perspective may be in order here.

The divisibility of atoms is made manifest, and their deconstruction occurs, when rarefied gases are subject to strong electric fields. The process could be realized and studied conveniently within Crookes tubes—that is, glass tubes evacuated to near vacuum, with electrodes at each end. As we understand it today, a voltage difference across the electrodes creates electric fields that will accelerate charged particles within the gas. Ionized atoms and electrons, always present at some small density owing to cosmic rays or thermal excitation, get boosted to high energy. They become projectiles, capable of breaking apart other atoms. The products include new charged particles, which are accelerated in turn, in a chain reaction that makes bottled lightning.

In 1897 J. J. Thomson made an epochal discovery. As just discussed, in a high-voltage Crookes tube the charged debris of atom-breaking collisions streams toward the electrodes. Some fast-moving positively charged ions collide with the lower-voltage electrode (the cathode). Negatively charged particles emerge from those impacts and stream toward the high-voltage electrode (the anode). Thomson studied these "cathode rays." He established that they consist of particles, each with the same quantity of electric charge and the same mass, regardless of what the cathode is made from. He went on to demonstrate that hot materials and radioactive materials emit particles with those same characteristics. Putting it together, Thomson deduced that there was a universal *subatomic* building block of matter. This is, of course, what we now call the electron.

The cathode rays have a simple universal character, being made of electrons. The positively charged *anode rays* are another story altogether. They are the remainders of atoms after one or more electrons get stripped away. Those remainders (we now know) consist of the atomic nuclei, together with a variable number of electrons. Thus anode rays come in many varieties and retain features that distinguish among different chemical substances. The pioneering work in Crookes tubes thereby posed a challenge as clear as it was grand: to continue the analysis of matter by understanding, as concretely as one understood electrons, those complementary components of atoms. The challenge was heightened when the mass of individual electrons was found to be only a tiny fraction (less than 1/1000) of the total mass of individual atoms.

Besides (and before) the discovery of electrons, the pioneering work on high-voltage electric discharges in Crookes tubes gave subatomic science another big gift: X-rays. Wilhelm Röntgen is generally credited for their discovery, as his systematic experimental studies of 1895—including a spectacular image of the bones of his wife's hand—brought the subject to an entirely new level. The possibility of high-frequency electromagnetic waves was already implicit in James Clerk Maxwell's 1861 synthesis (Maxwell's equations), and on that basis Hermann von Helmholtz mathematically anticipated the existence of a new type of highly penetrating radiation. Still, Röntgen's discovery had tremendous psychological impact: suddenly it was important and plausible to look for *weird* physical phenomena, purely experimentally.

Complacent faith in the near-closure of classical physics was no longer viable.

The strangeness and vividness of actual X-rays, and their potential for medical and scientific applications, inspired a surge of experimental activity. This exploration led to major progress on several fronts, including the discovery of the electron. But for our story the most important result was a chance discovery by Henri Becquerel, in 1896.

Becquerel was studying phosphorescence, that is, the ability of certain materials to absorb high-frequency electromagnetic energy, such as the ultraviolet part of sunlight or X-rays, and then to emit some of that energy as visible light.[1] Phosphorescence was, and still is, a convenient way to detect and monitor X-rays. Becquerel found, however, that the uranium salts he was studying would phosphoresce spontaneously and at a steady rate, without prior exposure to sunlight, X-rays, or any other energy source. Furthermore, he demonstrated that some of the spontaneous radiation from these salts is more penetrating than ordinary light, or even X-rays, being capable of passing through opaque paper or even metal sheets. Becquerel had discovered a fundamentally new behavior of matter: radioactivity.

Nineteenth-century theoretical physics was utterly unprepared for this fundamental instability of matter. Caught by surprise, it had no answers to the most basic empirical questions: What materials are radioactive? What, exactly, do they emit? Answers could come only from experiment.

In celebrated, heroic chemical work, Marie and Pierre Curie isolated elemental sources of radioactivity. A host of

1. Phosphorescence that ceases rapidly once the energy source is removed is called fluorescence, but I won't insist on that distinction, which is not fundamental.

investigators, with Rutherford at the forefront, got to work analyzing the rich, complex story of exactly who decays into whom, emitting what.[2]

Radioactive Transformations chronicles the progress made over the first decade following Becquerel's surprise. Read in that light, the achievement is astounding. Having appropriate humility, I will venture no further here into the particularities of radioactive transformations, deferring to Rutherford's text.

The theoretical physics of 1905 could not locate the significance of radioactivity in the grand scheme of things. With hindsight, we know that theoretical understanding in this realm could not progress very far without revolutionary insights from quantum mechanics and (special) relativity—ideas that in 1905 were just aborning. Indeed the most basic aspects of radioactivity—the spontaneity of decays, the integrity of atoms, and the dual particle-wave nature of gamma rays—feature characteristic *quantum mechanical* behaviors.

Radioactive decays are spontaneous, and external conditions do not affect their rate. Furthermore it is unpredictable which particular atomic nuclei will decay within a given interval of time; only the overall rate of decay, averaged over many nuclei, is fixed. Those features were already suggested in the earliest work on radioactivity, as duly emphasized by Rutherford. Cognizant of the difficulty of reconciling the apparent facts with conventional notions of causality and determinism, he raised the issue of whether some hidden subatomic structure with subtle long-term instabilities might be

2. I should note that this way of framing the issue, which now is so familiar and seems so natural that we can scarcely avoid projecting it onto the phenomena, was itself a major conceptual innovation: it is the "disintegration theory" of Rutherford and Soddy.

at work inside radioactive materials. Today most physicists have come to accept individual indeterminism within statistical predictability instead as a fundamental feature of the world. It is certainly a foundational principle of quantum theory.

Ironically, the occasional spontaneous decay of atoms highlights, by way of contrast, the profound *integrity* atoms ordinarily display. As Rutherford does not fail to note, the very possibility of chemistry and spectroscopy, which rely on all atoms of the same element displaying the same intricate behaviors wherever and whenever they are observed, belies the possibility of modeling their decay as gradual erosion followed by sudden collapse and disintegration. The integrity of atoms posed an insurmountable problem for classical physics. It inspired Niels Bohr's heretical introduction of "stationary states" in his atomic models of 1913, which initiated the quantum theory of matter.

It was natural, as Rutherford notes, to interpret one form of radioactive emission, the gamma rays, as electromagnetic pulses, that is, part of a continuum extending light and X-rays to still higher frequency. The gamma rays were often produced in association with rapidly accelerated electrons (beta rays), they were much more penetrating than the other common radiations (alpha and beta rays), and they were not deflected by magnetic fields. All these properties are consistent with expectations for high-frequency electromagnetic waves, whose existence and properties followed from Maxwell's equations. And yet the gamma rays seemed to be particles, not waves: they deposit their energy along straight paths. This situation, that equations for waves are associated with manifestations of particles, epitomizes another central, general feature of the quantum world.

These quantum features of the nuclear world leap out, as

experimental facts, in radioactivity. They were not ripe for interpretation, however, in the historical development. Their context was too unfamiliar and poorly understood. The very concept of "atomic nucleus" emerged only in 1913, and a reasonably coherent (though still crude) picture of atomic nuclei was achieved only in 1931. The rules of quantum physics were instead inferred, for the most part, from studies in more mature branches of physics, especially the thermodynamics of electromagnetic radiation (black-body formula) and atomic spectroscopy (Bohr atom), and—more remotely—William Rowan Hamilton's mathematical synthesis of particle mechanics and wave optics. What is remarkable, philosophically, is that the rules derived in those tame, domesticated contexts proved to apply also in the much more extreme, exotic context of nuclear (and later, subnuclear) transformations. Indeed, the most paradoxical elements of quantum theory are on display, stark and unadorned, in the simplest observations on radioactivity.

Rutherford's Silliman lectures for 1905 report a very substantial development of experimental nuclear physics, while 1905 is also, famously, the year of Albert Einstein's first relativity papers. As with quantum theory, a mature understanding of special relativity theory *might have* sped the development of nuclear physics, had such mature understanding been available. There is a widespread misconception that Einstein's special relativity, and specifically the mass-energy relation $E = mc^2$, ushered in the nuclear age. In reality the two fields, nuclear physics and relativity, developed in parallel and almost independently.

Radioactive Transformations touches on a fundamental issue that was in the air at the time, which really was close to the concerns that stimulated special relativity: the question of the origin of the electron's mass, and how that mass might be affected by its motion. On pages 10–11 we find:

> J. J. Thomson had shown in 1887 that a charged body in motion possessed electrical mass in virtue of its motion. . . .
>
> The moving charge acts as an electric current, and a magnetic field is generated round the body and moves with it. Magnetic energy is stored in the medium surrounding the charged body, which consequently behaves as if it had a greater apparent mass than when uncharged. This additional electric mass, according to the theory, should be constant for small speeds but should increase rapidly as the velocity of light is approached.
>
> Kaufmann found from his experiments that the apparent mass of the electron did increase with speed, and that the increase was rapid as the velocity of light was approached. . . .
>
> This was a very important result, for it indirectly offered a possible explanation of the origin of mass, which has always been such an enigma to science. If a charge of electricity in motion exactly simulates the properties of mechanical mass, it is possible that the mass of matter in general may be electric in origin, and may result from the movement of the electrons constituting the molecules of matter.

And on page 260:

> We thus arrive at the remarkable conclusion that the particles of the cathode stream and the β particles of radium are not matter at all in the ordinary sense, but disembodied electrical charges whose motion confers on them the properties of ordinary mass.

Einstein's work gave an alternative account of the apparent increase of mass (or, as we might say today, inertia) with velocity, and established the speed of light as the limiting speed.[3] The attractive idea that the electron's mass might be explained as field energy never proved very fruitful; indeed, modern renormalization theory discredits it. On the other hand, closely related ideas really do explain, in the framework of quantum chromodynamics, most of the mass of atomic nuclei!

Only *after* mature nuclear models had been formulated could the accurate formulas of relativistic mechanics, necessarily including the possibility of converting mass into energy, and conversely converting energy into mass, be of serious use. Their application has been very important and productive.

So much for context and background. Allowing *Radioactive Transformations* to speak for itself, my next task is to locate its achievement in the perspective of later developments. I shall now briefly review developments whose roots can be clearly discerned in *Radioactive Transformations.*

The key discovery leading to modern, successful atomic models was made by Hans Geiger and Ernest Marsden in 1911. Working in Rutherford's laboratory and following his suggestion, Geiger and Marsden studied the deflection of alpha particles, emitted in radioactive decay of radium, as they strike a gold foil. They discovered that a small but easily measurable fraction of those alpha particles are scattered through very large angles. Since 1907 alpha particles had been known, again through Rutherford's work, to be completely ionized helium atoms—that is, helium atoms lacking two electrons, which we *now* recognize as helium nuclei. The

3. It is now standard, when speaking of the mass of a particle, to refer to its rest mass. The velocity-dependent mass that figured in the early literature (in two varieties, longitudinal and transverse) caused ambiguity and confusion and has been abandoned. Of course the underlying empirical fact, that it becomes increasingly difficult to accelerate a particle as its speed approaches the speed of light, remains valid.

alpha particles have substantial inertia, so they can't be much deflected unless they encounter some stiff resistance; roughly speaking, unless they bounce from a small, very heavy object. Rutherford had not expected such large deflections to occur:

> It was quite the most incredible event that has ever happened to me in my life. It was almost as incredible as if you fired a 15-inch shell at a piece of tissue paper and it came back and hit you. On consideration, I realized that this scattering backward must be the result of a single collision, and when I made calculations I saw that it was impossible to get anything of that order of magnitude unless you took a system in which the greater part of the mass of the atom was concentrated in a minute nucleus. It was then that I had the idea of an atom with a minute massive center, carrying a charge.[4]

Rutherford proposed a definite, remarkably simple model that explains the observations. He proposed that within each atom there is a tiny nucleus, containing all of its positive charge and almost all of its mass. The remainder of the atom, according to Rutherford, consists of negatively charged electrons, dispersed over a much larger volume. He put this model to work, and validated it, by accounting *quantitatively* for the large-angle scattering. Extrapolating the standard formulas for electrical forces—that is, Coulomb's law, according to which the force is proportional to the product of charges and inversely proportional to the square of the separation—he calculated the number of alpha particles that will be scattered as a beam travels through material dotted

4. Quoted in David C. Cassidy, Gerald Holton, and Floyd James Rutherford, *Understanding Physics,* Harvard Project Physics (Birkhäuser, 2002), p. 632.

with point-like nuclei.[5] One can compare the results not only for overall rate but for the rate as a function of the angle of deflection. Rutherford's calculated rates, based on this simple atomic model, matched the observations.

This was an epochal result. It showed that the problem of understanding the atomic structure of matter could be divided conveniently into two parts. They correspond to what we now call atomic and nuclear physics.

The aim of atomic physics is to consider a heavy, positively charged nucleus as given, and then to determine how electrons are bound to it. Rutherford's explanation of the Geiger-Marsden result, based on electrical forces, also made it plausible that a known force law, namely the electric attraction of oppositely signed charged particles, could be extrapolated correctly down to subatomic distances.

It was natural to ask whether that force might be *all* that is required to build accurate atomic models.

That attractive idea, unfortunately, foundered on a very basic difficulty. Consider the simplest atom, hydrogen, with a single electron. According to classical mechanics and electromagnetism, as the electron orbits the nucleus it emits electromagnetic radiation, losing energy and spiraling in. There are no stable orbits.

Niels Bohr, then a young visiting scholar—the house theorist, we might say—with Rutherford in Manchester, cut through that Gordian knot in 1913. He kept the classical force law but boldly modified the rules of mechanics. He proposed that not all orbits are allowed but only a discrete subset of them, for which certain dynamic quantities are whole-number multiples of a universal constant, Planck's

5. This particular experiment put an upper limit of 3.4×10^{-14} meters on the nuclear size, which is about five times what we now know the size to be.

quantum.[6] Electrons can decay from higher-energy orbits into lower-energy ones, emitting electromagnetic radiation whose frequency, according to earlier ideas suggested by Max Planck and Einstein, should be proportional to the energy difference. In this way Bohr was able to account quantitatively for the spectral lines of hydrogen. Bohr's success validated Rutherford's basic picture of atoms and set the agenda for a generation of theoretical physicists: to ground Bohr's ad hoc rules into a logically coherent mathematical theory. It was in those struggles that modern quantum theory was forged.

The aim of nuclear physics is to understand what those inner cores of atoms, their nuclei, are made of, and to comprehend the laws that govern them. Here it becomes clear that electric forces will not suffice. Indeed the nuclei feature concentrated positive charge, which—if not overbalanced by other forces—will blow apart through electric repulsion.[7] New forces, unknown to classical physics, had to be at work.

Thus nuclear physics posed two challenges: the existential challenge, of identifying the ingredients of nuclei, and the dynamical challenge, of understanding the forces that those ingredients exert on one another.

The census of ingredients was settled in a few years, and rather simply. One ingredient was more or less obvious. The hydrogen nucleus is stable, is (apparently) indivisible, and carries one (positive) unit of electric charge. It is the lightest of all nuclei, and other light nuclei have masses that are close to whole-number multiples of its mass. So this *proton*—named

6. For experts: in Arnold Sommerfeld's crisp reformulation, the condition is that the action, integrated over an orbital period, is an integer multiple of Planck's constant.

7. The other force from classical physics, gravity, is negligibly small in nuclear physics.

by Rutherford—was one ingredient. The most economical assumption was that this is the only new ingredient. Nuclei might consist, like the atoms of which they are the core, of protons and electrons bound together, with powerful new short-range forces enabling much tighter binding.

In 1920 Rutherford proposed a refinement of that idea. Both the proposal and the reasoning behind it proved prescient:

> Under some conditions, however, it may be possible for an electron to combine much more closely with the H nucleus, forming a kind of neutral doublet. Such an atom would have very novel properties. Its external field would be practically zero, except very close to the nucleus, and in consequence it should be able to move freely through matter. Its presence would probably be difficult to detect by the spectroscope, and it may be impossible to contain it in a sealed vessel. On the other hand, it should enter readily the structure of atoms, and may either unite with the nucleus or be disintegrated by its intense field.
>
> The existence of such atoms seems almost necessary to explain the building up of the nuclei of heavy elements; for unless we suppose the production of charged particles of very high velocities it is difficult to see how any positively charged particle can reach the nucleus of a heavy atom against its intense repulsive field.

The properties Rutherford imputes to his "neutral doublet" are very close indeed to the properties of the neutron, discovered by James Chadwick in 1931.[8] Rutherford was led

8. Chadwick quoted the above passage in his 1935 lecture accepting the Nobel Prize for his discovery of the neutron.

to his idea by the problem of understanding how, by physical means, heavy nuclei could ever have been assembled. The difficulty is that powerful electric repulsion acts between nuclei and makes them difficult to bring together. Even if new attractive forces come into play at short distances and are capable of fusing the nuclei once they are brought together, first that repulsive barrier must be overcome. Rutherford envisaged a clever way around the problem: his neutral doublets would feel no repulsion, and so they might form a delivery system to sneak additional protons (together with tightly bound electrons) into nuclei. Once inside, the electrons might be stripped from the doublets and expelled, to make beta radiation. This way of building up heavy nuclei is essentially what occurs in supernova explosions, and also in nuclear reactors and bombs.

Yet there was never any evidence for the sort of powerful new forces between electrons and protons that Rutherford's "neutral doublet" required. In nuclear physics the neutron stands on its own, an independent ingredient as fundamental as the proton.[9]

The experimental discovery of the neutron, an electrically neutral particle only slightly heavier than a proton, was a big advance, because it allowed a simple yet useful picture of what nuclei are, namely, collections of protons and neutrons bound together. With that picture, many observed facts fell into place.

Nuclei of different elements differ in the number of protons they contain. That number determines the electric charge of the nucleus, which in turn controls its interaction with the surrounding electrons in an atom. Those surrounding electrons, in turn, control the atom's chemical properties.

9. Today we know that both protons and neutrons are made from quarks and gluons, according to very similar body plans.

Two nuclei can have the same atomic number, that is to say the same number of protons, but different numbers of neutrons. Such nuclei are called isotopes. Atoms containing isotopic nuclei will have the same chemical properties but will differ in weight. They also differ in stability; for example, different isotopes of uranium exhibit drastically different levels of radioactivity.

The total mass of a nucleus is approximately—but only approximately—equal to the sum of masses of the protons and neutrons that make it up. This is a most profound fact, which marked the emergence of mature nuclear physics. The basic idea is insightful, beautiful, and nicely captured in three simple equations.

For a quantity of protons, Z, and N neutrons we have the total mass and rest-energy

$$M_{\text{constituents}} = Zm_{\text{proton}} + Nm_{\text{neutron}}$$
$$E_{\text{constituents}} = (Zm_{\text{proton}} + Nm_{\text{neutron}})c^2. \tag{1}$$

In the nucleus, there is additional energy associated with the interactions, so we have

$$E_{\text{nucleus}} = (Zm_{\text{proton}} + Nm_{\text{neutron}})c^2 + E_{\text{interactions}}. \tag{2}$$

Dividing by c^2, we have for the mass of the nucleus

$$M_{\text{nucleus}} = (Zm_{\text{proton}} + Nm_{\text{neutron}}) + E_{\text{interactions}}/c^2 = M_{\text{constituents}} + E_{\text{interactions}}/c^2. \tag{3}$$

Thus the difference between the measured mass of a nucleus and the total mass of its constituents, which are both measurable quantities, is $E_{\text{interactions}}/c^2$. It is known as the mass defect of the nucleus.[10]

10. For experts: since the interactions are primarily attractive, $E_{\text{interactions}}/c^2$ is negative.

The fact that the mass defect is much smaller than the naive "constituent counting" mass—it never reaches more than five percent, for any nucleus—shows that the interactions of protons and neutrons within nuclei, though powerful, are not so strong as to challenge their integrity as mass-units. In this quantitative sense, the nucleus *is* a collection of definite numbers of protons and neutrons.

Clarity as to the constitution of nuclei, as just described, also brought order into the description of their transformations. At first the variety of nuclear transformations seemed bewildering—*Radioactive Transformations* testifies to that! Yet the naturally occurring, spontaneous processes of radioactivity only initiated the subject. Additional transformations were observed to result from impacts of energetic radioactive emissions, especially alpha particles, on target foils.

One kind of transformation involves changes in the disposition of protons and neutrons, without any change in their numbers. An example is the original nuclear reaction (4) studied by Rutherford in 1917.

$$^{14}\mathrm{N} + \alpha \rightarrow {}^{17}\mathrm{O} + p \qquad (4)$$

Here the superscripts indicate the total number of protons plus neutrons in the nucleus. Rutherford discovered that upon bombarding nitrogen with alpha particles, one sometimes observed oxygen nuclei and protons as products. Counting up the protons and neutrons: on the left-hand side we have seven protons and seven neutrons in nitrogen (^{14}N) and two protons and two neutrons in the alpha particle; on the right-hand side we have eight protons and nine neutrons in oxygen (^{17}O) and an additional proton. So on each side we have nine protons and nine neutrons. In the reaction these particles have been rearranged but neither created nor destroyed. Most radioactive decays involving alpha emission are of this kind. The parent nucleus turns into a different nucleus, which has two protons and two neutrons fewer, together with the alpha particle.

The other kind of transformation involves neutrons converting into protons (or vice versa). The prototype of this kind of transformation is the decay of a free neutron into a proton, an electron, and an antineutrino:

$$n \rightarrow p + e + \overline{\nu} \qquad (5)$$

When it occurs inside a nucleus, this transformation leads to emission of an electron, or beta ray. Proper discussion of the antineutrino would involve us in a long digression; suffice it to say that it is a neutral particle whose interactions with matter are very feeble, so that in the early experiments it was not detected at all.

The reverse conversion is not very important in radioactivity, but it is the central process powering stars:[11]

$$p \rightarrow n + \overline{e} + \nu \qquad (6)$$

Here $\overline{e}$ is an antielectron (called a positron), and ν is a neutrino. Conversion of protons into neutrons cannot happen for isolated protons, since the neutron is heavier, but it can and does happen in nuclear environments, where more favorable interaction energy for the neutron can compensate for its unfavorable rest-energy. Decays involving conversions between neutrons and protons are invariably slow, and reactions involving such conversions are invariably rare.

All the observed transformations of nuclei, including the

11. There are a few examples of nuclei that decay through the interesting process of electron capture, following the schema

$$p + e \rightarrow n + \nu. \qquad (7)$$

In this process, the nucleus "captures" one of the surrounding electrons from its atomic cloud.

many processes recorded in *Radioactive Transformations,* are of these two basic kinds: rearrangements of protons and neutrons, or single neutron and proton interconversions. The latter are accompanied by emission of electrons, neutrinos, or their antiparticles.

Mature understanding of what nuclei are made of, and how they transform, led to several important applications.

Stellar Nucleosynthesis. Nuclear transformations provide the primary source of energy for stars, including our Sun. In normal ("main sequence") stars the dominant process is fusion of hydrogen into helium. It proceeds through a number of intermediate steps, including two proton-to-neutron conversions. The net result is

$$4\,{}^{1}\mathrm{H}\,(=4p) \rightarrow {}^{4}\mathrm{He}\,(=\alpha) + 2\,\bar{e} + 2\nu. \qquad (8)$$

The hydrogen nucleus ^{1}H is the proton, and the helium nucleus ^{4}He is none other than the alpha particle. ^{4}He contains two protons and two neutrons. It is a particularly stable nucleus, with a large mass defect. Owing to that mass defect, the reaction (8) liberates energy. This is the energy that powers our Sun.

It is significant that the stellar fusion of hydrogen into helium requires two proton-to-neutron conversions, which are rare events. That is why stars can stay alight for billions of years, supporting a slow but steady burn.

When a star runs out of hydrogen fuel, it starts to contract, raising its temperature. Eventually the temperature rises so high that helium nuclei fuse. That process requires a higher temperature than hydrogen fusion, in order to overcome the stronger barrier of electric repulsion facing the more highly charged helium nuclei. The dominant fusion, $3\,{}^{4}\mathrm{He} \rightarrow {}^{12}\mathrm{C}$, involves only rearrangement of protons and neutrons, not any conversion, so it occurs relatively rapidly.

As helium is exhausted, there can be additional rounds of fusion at still higher temperatures, but eventually the star runs out of fuel. What happens at that point depends primarily on the size of the star. Stars like our Sun settle down into white dwarfs. But the collapse of significantly more massive stars is catastrophic. The gravitational energy unleashed by the collapse heats the infalling matter to extraordinarily high temperatures, causing it to explode as a supernova. (This is a rough sketch of one class of supernovae, the so-called type II supernovae. Type I supernovae arise when matter from a companion star accretes onto a white dwarf, until the bloated dwarf collapses under its own weight.)

Supernova explosions populate the interstellar medium with heavy nuclei. Some of these (for example, ^{12}C nuclei) are simply the ashes of earlier fusions; others, including all those heavier than iron, are produced in the explosion itself. The path to these heavier nuclei is basically the one that Rutherford conceived in 1920: the exploding matter contains many free neutrons. Facing no barrier from electrical repulsion, neutrons readily enter and accumulate on ambient nuclei, allowing them to grow.

There is a richly detailed theory of the astrophysical processes that produce a wide variety of nuclei, starting from a mixture of hydrogen and helium.[12] It gives an excellent account of the relative abundances of nuclei as observed in Nature. Thus we obtain, from fundamental nuclear physics and astrophysics, a validated quantitative explanation of the origin of the chemical elements, and also of their isotopes.

The Origin of Radioactivity. If radioactivity is a process of decay, leading from active (unstable) nuclei to inert products, how—or why—did it ever begin? The theory of stellar

12. A few of the lightest nuclei were produced primarily in the big bang. See below.

nucleosynthesis, with the final liberation of synthesized heavy elements in supernova explosions, provides a poetic, historically resonant explanation of radioactivity's origin.

In the violent last throes of its disintegration, as a star explodes, vast quantities of energy are pumped into the escaping matter. In that environment barriers that normally separate different nuclei, or nuclei and neutrons, are readily overcome, and the particles fuse. As these supercharged, unstable fusions settle down, some of their energy gets hung up, locked into forms that can leak out only very slowly. This is the song of phosphorescence, transposed from atomic into nuclear keys. Radioactivity is the phosphorescence of stardust.

Big Bang Nucleosynthesis. In the framework of big bang cosmology, it is possible to carry the analysis of nuclear origins back to the beginning.

In the earliest moments of the big bang, all the material of the universe was at such high temperature, and so dense, that no complex nuclei could persist. Violent impacts with surrounding matter shattered any incipient nuclei back into protons and neutrons. But as the universe expanded and cooled, some more-complex nuclei were created.

The early cosmological environment was quite different from conditions in stars. The density was lower, but the temperatures, at least initially, were higher. The mixture was more neutron-rich, but the time available for reactions, before the expanding medium became too tenuous, was more limited.

It is straightforward to work out the consequences of big bang nuclear synthesis. Its predicted result is mostly ^{1}H, but also a substantial fraction of ^{4}He, significant ^{2}H (deuterium) and ^{3}He, and a trace of ^{7}Li.[13] The observed abundances of these nuclei are in accord with the predictions.

13. The predicted ^{7}Li abundance is small indeed, but because this nucleus is even harder to produce in stellar environments, the big bang contribution must—and does—account for the observed abundance.

Averaged over the universe as a whole, the dominant nuclei are ^1H and ^4He, reflecting big bang residues; with the above-noted exceptions, other nuclear species are products of stellar burning and supernova explosions.[14]

Radioactivity and Chronology. In the nineteenth century there was a great scientific controversy about the age of the Earth. Physical arguments seemed to point to a relatively modest age, around 20 million years, which seemed inadequate to geologists and evolutionary biologists.

The two main physical arguments concerned the heat of the Sun and the heat of the Earth.

One can calculate the rate at which the Sun is presently radiating energy. To account for that energy, Lord Kelvin proposed that energy from the original accretion of the solar material might have been accumulated and then only gradually released. This would suffice for a few tens of millions of years of present-day solar output. He also examined other possibilities but found none more promising.

As we've discussed, the Sun has another energy source, unknown to Kelvin: nuclear fusion. That source can comfortably support several billion years of stable radiation.

Mining operations reveal that the temperature of Earth is higher in its interior. Thus the Earth is radiating internal heat. Where does the energy come from? Again, the most promising suggestion seemed to be that the gravitational energy of formation was gradually leaking away. This process, Kelvin estimated, would support the present rate of cooling for a few tens of millions of years—suggestively close to the calculated solar age. Rutherford reviews this issue on pages 213–217, where he emphasizes that radioactive decays supply a possible additional heat source. Today we know that he was

14. Actually there is one more complication: a few rare nuclei arise as byproducts of cosmic ray spallation, that is, as fragments of heavy nuclei struck by energetic cosmic ray protons.

on the right track, although in 1905 he could not be specific as to details. Earth is heated by radioactive decays in its interior, especially decays of potassium ^{40}K. This decay accounts for the high temperatures underground and provides the energy that drives plate tectonics.

Besides resolving these controversies of principle, radioactivity furnishes a powerful constructive tool for dating materials. Here the pioneering work was Rutherford's 1905 estimate of the age of the Earth, or more precisely of ancient minerals, described on pages 187–191. His methodology launched a fruitful, wide-ranging field of science.

Nuclear Technology. Scientific understanding of the principles of nuclear transformations has enabled new technologies. Most awesome, both in present achievement and in future potential, is the access to energy that nuclear processes offer, on scales that dwarf conventional chemical processes. The power of nuclear weapons is all too familiar. Nuclear reactors will become attractive power generators, free of carbon emissions, if issues surrounding disposal of their (quite different!) waste products can be convincingly addressed. Controlled fusion has been an alluring dream for decades. Many technical difficulties have been overcome, and controlled fusion itself is now routine, though large-scale energy production at economically competitive rates remains futuristic. Progress in these areas could have very large leverage indeed; one can easily fantasize about world-historic innovations.

Quite a different nuclear technology, which has made a large and unambiguously positive contribution to human welfare, is nuclear medicine. This area has many facets, but I'll mention just one here, for its historical resonance. By attaching radioactive nuclei to biologically active materials, one can deliver those nuclei to places of interest in a human body. For example, one can attach radioactive nuclei to substances readily taken up by cancer cells. Then when the radioactive nuclei decay, they reveal where the cancer is. In this way we can now take photographs "from the inside out," as X-rays enabled us to take them "from the outside in."

The mature nuclear physics just described, to which Rutherford contributed so decisively, is evidently a superb scientific achievement, with wide-ranging ramifications. It provides a rough but serviceable picture of atomic nuclei, which can be used to organize a wealth of data and enable impressive applications in astrophysics, cosmology, and technology.

Yet to physicists, semi-empirical nuclear physics remained, manifestly, an unfinished product. It achieved its successes by codifying a wealth of experimental facts in simple semi-empirical models, finessing ignorance about the fundamental forces.

I shall now sketch later developments in the subject area, which go well beyond anything Rutherford could envisage.

The experimental study of nuclear forces soon led in unanticipated directions. The main method of investigation was the scattering experiment. Though details of implementation and interpretation are often difficult and complicated—one is dealing with *very* small and unfamiliar objects, after all—the basic concept of scattering experiments is straightforward. It is, in fact, the same concept that guided the historic Geiger-Marsden experiment.

To investigate, say, the force between protons, one can shoot beams of protons at other protons (for example, a hydrogen target) and investigate their deflection. Then from the rates at which deflections through different angles occur, one can try to infer the underlying force. One can use beams of protons with different energy, and with the protons spinning in different directions, to enrich the analysis.

Experiments of this kind soon revealed that the forces between protons and neutrons do not obey any simple equation. They depend not only on distance but also on velocity and spin, in complicated ways.

More profoundly, scattering experiments soon undermined the notion that protons and neutrons are simple particles, or that any sort of traditional "force" between them could do justice to the reality of nuclear physics. For when high-energy protons collide with other protons, the typical result is not merely a deflection but the production of new particles.

In fact a whole world of new particles was discovered in this way, π, ρ, K, η, ω, K*, and φ mesons and Λ, Σ, Ξ, Δ, Ω, Σ*, Ξ*, and Ω* baryons being among the lightest and most accessible. The details are fascinating to experts, but only a few broad features will concern us here.

Protons and neutrons are the prototype of baryons, and all baryons share several properties. They all feel strong short-range forces in one another's presence, or in the presence of mesons, and (for experts) they are all fermions.

The most profound feature of baryons is their conservation law. Previously, we saw that nuclear transformations include processes where neutrons convert into protons, or vice versa. But the total number of protons plus neutrons remains the same, or (we say) is conserved, despite such transformations. In the processes observed at higher energies, protons can convert into other baryons, not only neutrons. Yet the total number of baryons, adding up all types, is conserved in all processes.[15]

Mesons also share common properties. They all feel strong short-range forces in one another's presence, or in the presence of baryons, and (for experts) they are all bosons. There is not a conservation law for mesons.

15. Although no violation of the law of baryon number conservation has ever been detected experimentally, there are good reasons to suspect that it is not strictly exact. See below.

Very roughly speaking, we can say that baryons resemble the traditional notion of material particles, while mesons can be considered force-mediating particles, or field-quanta, analogous to photons. (But the photon itself is not considered a meson, because it does not exhibit the strong short-range interactions characteristic of mesons.)

Mesons and baryons, collectively, are known as *hadrons.* Aside from the proton and neutron, hadrons are all highly unstable particles, which decay in a small fraction of a second. Nevertheless they exist, and they can be observed and studied in considerable detail.

Thus a major suggestion from postnuclear physics is that protons and neutrons are not fundamental particles but just two members among a much larger family of closely related particles, the hadrons. The complexity of proton-proton forces conveys the same suggestion. The complexity of proton-electron and proton-photon forces, revealed in parallel high-energy studies, is even more convincing: because electrons and photons *are* simple elementary particles, whose fundamental interactions are known reliably,[16] the complexity of their interactions with protons must be ascribed to complex structure within the protons.

According to the quark model, baryons are bound states of three more-fundamental entities: quarks. Quarks come in six "flavors": up *u,* down *d,* strange *s,* charm *c,* bottom *b,* and top *t.* Of these only *u* and *d* quarks appear in protons and neutrons, while only *u, d,* and *s* appear in the low-mass baryons and mesons enumerated above (and many others); the heavy and highly unstable *c, b,* and *t* quarks are relatively recent additions.

The quark model was a major step in organizing the the-

16. The interactions of photons and electrons among themselves, and with nuclei at low energies, can be accurately described using simple, elegant equations.

ory of the hadronic world. It provides a picture of hadrons analogous, in its explanatory power, to Bohr's model of atoms—that is, correct in spirit and historically important, but logically incomplete and only semi-mathematical.

How do three kinds of quarks generate hundreds of different baryons? The point is that a given trio of quarks, say u, u, d, can exist in many discretely different states of motion (analogous to Bohr's quantized orbits for electrons). These different states will have different energies, and therefore—using $m = E/c^2$—different masses. Thus they appear, operationally, as different particles. In this way, we find that many different particles can all correspond to the same underlying material structure, captured in different states of internal motion.

Similarly, the quark model postulates that mesons are bound states of a quark and an antiquark. A given quark-antiquark pair, say $u\,\bar{d}$, in various states of motion, generates many different mesons.

The quark model gives a plausible explanation for the complexity of hadronic forces, as well. Even if quarks have simple interactions, bound states containing three quarks, or a quark and an antiquark, offer many opportunities for cross talk and cancellations. Indeed, it is for reasons like this that chemistry (that is, interactions of atoms) is extremely complicated even though the underlying forces between individual electrons are extremely simple.

The quark model as such, however, neither relied upon nor provided a specific theory of the forces among quarks. Maxwell's equations for electrodynamics, Isaac Newton's and then Einstein's equations for gravity, and Erwin Schrödinger's and then Paul Dirac's equations for atomic physics set standards for beauty and accuracy that the equations of postnuclear physics, for several decades, could not approach.

The decisive breakthrough came in 1973, with the discovery of asymptotic freedom by David Gross and myself, and independently by David Politzer, and with the formulation of quantum chromodynamics (QCD) by Gross and me. A proper description of that work would necessarily distort the balance of this foreword, since it brings in several difficult new concepts, so I'll happily refer you to my Nobel lecture.[17] Here I'll only describe, in a general way, four major consequences that tie in with our present themes.

Quantum chromodynamics provides the sought-for beautiful, accurate equations governing the strong force. The structure of the QCD equations is similar to the structure of Maxwell's equations, but they are both more complex and more symmetrical.[18] (Metaphorically speaking, the equations of QCD are to Maxwell's equations as an icosahedron is to a triangle.) The equations of QCD provide the foundation for nuclear physics, in principle, but that application is two steps removed from the simple basics. First the forces among quarks and gluons bind them into protons and neutrons; then—as our discussion of the quark model suggested—complicated multiparticle interactions come into play when protons and neutrons influence one another. There has been remarkable progress, involving heavy use of supercomputers, in computing the structure of protons and neutrons from the equations of QCD, but accurate calculation of nuclear forces still lies in the future.

A major consequence of QCD is that there should exist, in addition to quarks, eight color gluons. These gluons play the same role in QCD as the photon plays in quantum electrodynamics (also known as QED).

Although QCD predicts that quarks and gluons cannot exist as isolated particles but are always "confined" within

17. "Asymptotic Freedom: From Paradox to Paradigm," in *Les Prix Nobel 2004* (Stockholm: Almqvist & Wiesell International, 2004), pp. 100–124.

18. C. N. Yang and R. Mills discovered the mathematical possibility of generalizing Maxwell's equations to embody larger symmetry, in 1956.

bound states such as baryons and mesons, nevertheless they are experimentally observable, in quite a direct way. In very high energy processes, one observes the emission of *jets.* Jets consist of several hadrons, all moving rapidly in the same direction. According to QCD, jets are the residue from the emission of a quark, antiquark, or gluon, observed after the original quark (or antiquark or gluon) has radiated additional gluons and quark-antiquark pairs, which self-organize into hadrons. The total energy and momentum of the jet reflect the energy and momentum of the quark (or antiquark or gluon) that triggered its formation, since energy and momentum are conserved. Thus by measuring jets, one can reconstruct quarks, antiquarks, and gluons, and compare their properties with those predicted by the equations of QCD. This is a much more direct and easier use of the equations than calculating nuclear forces, and has allowed the theory to be tested quantitatively in great detail.

Quantum chromodynamics gives a compelling account of the origin of most of the mass of protons and neutrons.

In QCD the proton is rather a more complex object than envisaged in the quark model. In addition to the trio *uud* of quarks that the quark model posited, protons contain additional quark-antiquark pairs and multitudes of gluons, coming to be and passing away in a dynamic equilibrium. But the crucial point is that the particles that make up a proton[19]—*u* and *d* quarks, their antiquarks, and gluons—are particles whose mass is quite small compared with the mass of the proton they build up.

So where does the proton's mass come from, if not from the mass of its constituents? There is energy associated with the internal motion of the quarks and gluons, even when the

19. There is also a small admixture of $s\bar{s}$ pairs, not important for this discussion.

proton as a whole is at rest. Let us call that energy E. That energy is concentrated in a small region of space, and seen from afar it looks like a particle (namely, a proton). According to special relativity, the localized energy E has the inertia of a mass $m = E/c^2$. And that is the origin of the proton's mass!

This account of the proton's mass, from QCD, is reminiscent of early ideas about the origin of the electron's mass from its electromagnetic field energy, which we discussed earlier, but it has the virtues of being precisely formulated and provably correct. Supercomputer calculations, working directly from the equations of QCD, give accurate quantitative results for the masses of hadrons, notably including protons and neutrons.

Another way of stating the result, in a language we used earlier, is to say that the mass of the proton is (almost) *entirely* its mass defect!

Quantum chromodynamics governs the basic dynamics that builds protons, neutrons, and the other hadrons out of quarks and gluons, and the forces that bind together nuclei—the so-called strong force. Quantum electrodynamics, including notably the electric repulsion between protons, modulates that dynamics.

Neither of those two great theories, however, incorporates processes whereby protons and neutrons interconvert. Such processes are associated with much weaker forces—that's why proton and neutron interconversion is slow (for decays) and rare (for reactions)—but since they bring in essentially new possibilities, they are both qualitatively important and readily detected. Indeed, they are responsible for many forms of radioactivity, and they play a crucial role in stellar energy generation.

To account for those rare conversion phenomena, physicists were led to postulate a fourth force, in addition to gravity, electromagnetism, and the strong force. This new addition, which completes our current picture of physics—the Standard Model—is called the weak force.

Postnuclear explorations in basic particle physics, based on observations of cosmic rays and work at accelerators, revealed that the "weak" force is not just a curious anomaly but rather a cluster of universal phenomena, encompassing a host of transformations and interactions. As with QCD, a proper description of the weak force would distort this foreword, so I'll confine myself to four brief comments on results that are relevant to my earlier themes or that lead into my upcoming conclusion.

First, at a fundamental level, the weak force isn't all that weak. In experiments that explore ultra-high energies, or (equivalently) that probe interactions at subnuclear distances, smaller than 10^{-16} centimeters, the weak interaction is seen to act more powerfully than electromagnetism.

Second, the equations governing the weak force are strikingly similar, in their mathematical form, to the equations of QCD, which in turn are generalizations of the equations of QED, namely Maxwell's equations.

Third, since protons and neutrons are complex composites of simpler, more-basic quarks and gluons, we should track proton-to-neutron interconversions to their more-basic source. The deep structures underlying the conversions (5, 6) are the quark processes

$$u \rightarrow d + \bar{e} + \nu \quad (9)$$

$$d \rightarrow u + e + \bar{\nu}. \quad (10)$$

Since the neutron differs from the proton by substitution of a d quark for a u quark, (9) induces (6) and (10) induces (5).

Finally, although it mediates conversions of one kind of

quark into another, the weak force, like the strong and electromagnetic forces and gravity, conserves the total number of quarks. This rule states the refinement, to the quark level, of the law of baryon number conservation.[20]

Since the strong, electromagnetic, and weak forces are governed by equations with very similar mathematical structures, it is natural to speculate that they are merely different aspects of a single more-general force. This is a concrete, modern form of the quest for a unified field theory.

This sort of speculation can be carried quite far. It can explain regularities in the patterns of particle interactions that the Standard Model ascribes to coincidence. Most impressively, it accounts quantitatively for the differences in strengths among the strong, weak, and electromagnetic interactions. Again, a proper description of these developments would lead us far afield, but one observation will serve as an appropriate conclusion to this foreword.

The unified theories necessarily include a wider class of transformations than occur in the separate parts of the Standard Model. Upon putting quarks, electrons, neutrinos, and their antiparticles on an equal footing, for example, we are led as a generalization of (9) to introduce the process (11) whereby an up quark converts into up and down antiquarks and an antielectron (= positron):

$$u \rightarrow \overline{u} + \overline{d} + \overline{e} \qquad (11)$$

This process (11) introduces a qualitatively new effect that is not present in the Standard Model: it violates the law of con-

20. Subtle effects in the weak interaction, and possibly in gravity, lead to violations of quark number conservation. Those effects are absurdly small and unobservable today but may have been significant in the earliest moments of the big bang.

servation of quarks, and the closely associated law of baryon number conservation. At the level of protons (and, therefore, nuclei), it leads to the possibility that protons can decay, into pi mesons and positrons:

$$p \rightarrow \pi^0 + \bar{e} \tag{12}$$

This process, were it possible, would undercut the long-term stability of all familiar forms of ordinary matter.

Decays based on the process (12) would represent a new form of radioactivity. With his preternatural instinct for Nature's dispositions, Rutherford pointed toward such a possibility in the final paragraph of *Radioactive Transformations.* Inspired by modern unified field theories, experimentalists have gone to great lengths in searching for it. So far the result has been negative, and therefore the rate of such decays must be quite small. The unified theories suggest rates that are not far beyond current limits. The search continues.

Frank Wilczek
MASSACHUSETTS INSTITUTE OF TECHNOLOGY
January 2012

PREFACE

The present work contains the subject matter of eleven lectures delivered under the Silliman Foundation at Yale University, March, 1905.

I chose as the subject of my lectures the most recent and at the same time the most interesting development of Radioactivity, namely the transformations which are continuously taking place in radioactive matter. While dealing fully with this aspect of the subject, it was necessary for clearness to give some account of radioactive phenomena in general, although with much less completeness than in my previous book on Radioactivity.

In arranging the chapters of the present volume, the order in which the subject was dealt with in the lectures has been closely followed, but as our knowledge of the subject is increasing so rapidly, I have thought it desirable to incorporate the results of the many important investigations which have been made since the lectures were delivered. This is especially the case in the chapter dealing with the α rays, to which much attention has been devoted in the past year on account of the important part they play in radioactive transformations.

I am much indebted to my colleagues Professor Harkness and Professor Brown for the great care and trouble they have taken in the correction of the proofs and for many useful suggestions.

E. RUTHERFORD.

McGill University,
Montreal, June 4, 1906.

RADIOACTIVE TRANSFORMATIONS

CHAPTER I

HISTORICAL INTRODUCTION

The last decade has been a very fruitful period in physical science, and discoveries of the most striking interest and importance have followed one another in rapid succession. Although the additions to our knowledge have come from investigations in very different fields, yet a close examination shows that they are all intimately related, and each discovery has supplied the necessary stimulus and suggestion to serve as a starting point for the next advance.

The march of discovery has been so rapid that it has been difficult even for those directly engaged in the investigations to grasp at once the full significance of the facts that have been brought to light. Especially has this been the case in the field of radioactivity, where the phenomena observed have been so complicated and the laws controlling them so unusual that it has been necessary to introduce conceptions of a novel character for their explanation.

The starting point of this epoch in physical science was the discovery by Röntgen of the X-rays in 1895 and the experiments of Lenard on the cathode rays. The extraordinary properties of the X-rays at once focussed the attention of the scientific world, and led to a series of investigations whose object was not only to examine the properties of the rays themselves, but to disclose their real nature and origin.

The latter problem led to a much closer investigation of the nature of the cathode rays produced in a vacuum tube, for these rays were seen to be in some way intimately connected with the emission of X-rays. J. J. Thomson in 1897 finally succeeded in proving definitely that the cathode rays consisted of a stream of particles moving with great velocities and carrying negative charges of electricity. These particles had an apparent mass only about $\frac{1}{1000}$ that of the hydrogen atom, and were therefore the smallest bodies known to science. These "corpuscles," or "electrons," as they have been termed, are apparently a constituent of all matter, and are believed to be the ultimate parts of which the atom is composed.

This electronic hypothesis has been extremely fertile, and has already greatly changed — or rather extended — previous conceptions of the constitution of matter. It has opened up wide fields of investigation in many departments of physical science, and has provided science with a microscope, so to speak, with which to examine the structure of the atom of the chemist. J. J. Thomson has examined mathematically the stability of atoms composed of a number of whirling electrons, and has shown that these model atoms imitate in a remarkable way some of the more fundamental properties of the chemical atom.

The proof of the corpuscular character of the cathode rays at once indicated the probable explanation of the origin and nature of the X-rays. Stokes, J. J. Thomson, and Weichert independently suggested that the cathode rays were the parents of the X-rays. The sudden stoppage of the electrons in the cathode stream causes an intense electromagnetic disturbance which travels out from the point of impact with the velocity of light. On this point of view, the X-rays consist of a number of disconnected pulses, following one another in rapid succession but without any definite order. They are akin in some respects to very short waves of ultra-violet light, but differ from them in the lack of periodicity of the pulses. The penetrating power of the rays, and the absence of any direct reflection, refraction, or polarization, were consequences of this theory if the breadth of the pulse was short compared with the diameter of the atom.

An admirable and simple account of the nature and properties of such pulses has been given by J. J. Thomson[1] in the Silliman lectures of 1903.

[1] J. J. Thomson: Electricity and Matter (Scribner, New York, 1904).

In the meantime, another remarkable property of the X-rays had been closely examined. The passage of the X-rays through a gas imparts to it a new power of rapidly discharging an electrified body. This was satisfactorily explained on the hypothesis that the rays produced a number of positively and negatively charged carriers or ions in the electrically neutral gas.[1] The development of this subject proceeded along two distinct lines, one electrical and the other optical. C. T. R. Wilson[2] found that under certain conditions the ions produced in the gas by X-rays become nuclei for the condensation of water upon them. Each ion thus becomes the centre of a visible charged globule of water which moves in an electric field. Experiments of this character verified in a remarkable way the fundamental correctness of the ionization theory, and clearly brought out the discontinuous or atomic structure of the carriers of the electric charges.

As a result of researches on diffusion of the ions in gases, Townsend[3] deduced the important fact that the charge carried by a gaseous ion was the same in all cases, and equal to the charge carried by the hydrogen atom in the electrolysis of water. By a combination of the electrical and optical methods, J. J. Thomson[4] found the actual value of the charge carried by an ion.

The determination of this important physical unit at once allows us to count the number of ions present in any volume of air acted upon by an ionizing agent. In addition to this, it provides the most accurate deduction yet made of the number of molecules present in unit volume of any gas at standard pressure and temperature. This number, which is based purely on experimental data, will be seen to be of the greatest value in calculating the magnitudes of various quantities in the subject of radioactivity.

The ionization theory of gases was successfully applied to

[1] J. J. Thomson and E. Rutherford: Phil. Mag., Nov., 1896.

[2] C. T. R. Wilson: Phil. Trans., p. 265, 1897; p. 403, 1899; p. 289, 1900.

[3] Townsend: Phil. Trans. A, p. 129, 1899.

[4] J. J. Thomson: Phil. Mag., Dec., 1898; March, 1903.

account for the conductivity of flames and heated vapors and to unravel the complicated phenomena observed in the discharge of electricity through a vacuum tube. This fascinating and far-reaching field of physical inquiry owes its inception and much of its development to Prof. J. J. Thomson and his students at the Cavendish Laboratory, Cambridge.

On the theoretical side the possibilities of an ionic or electronic theory of matter had been recognized long before the experimental evidence was forthcoming. The most notable exponents of this school were Lorentz and Larmor, who developed their theories to account, among other things, for the mechanism of radiation. The discovery by Zeeman of the action of a magnetic field in displacing the spectral lines afforded a strong confirmation of the general theory, for the experimental results observed were in large part predicted by the theory of Lorentz. In addition it was deduced that the ion, whose movements gave rise to the radiation, had a mass of about the same small value as the corpuscle of J. J. Thomson observed in a vacuum tube. Results of this character at once extended the range of ionic theories to matter in general, and though much still remains to be done, the electronic theory has already proved of great value in elucidating the connection between some of the most recondite physical phenomena.

The movement set on foot by the discovery of Röntgen had even more important consequences in another very unexpected direction. Immediately after the discovery of the X-rays, it was thought that the emission of these rays was in some way connected with the phosphorescence set up by the cathode rays on the walls of a vacuum tube.

It occurred to several scientists that natural bodies which phosphoresced under the influence of light might possess the property of emitting a penetrating type of radiation similar to X-rays. We now know that this speculation had no secure basis in fact, but it provided the stimulus for the investigation of the properties of substances in this special direction and soon led to a discovery of far-reaching importance.

M. Henri Becquerel,[1] a most distinguished French physicist, in pursuance of this idea exposed amongst other substances a phosphorescent uranium compound — the double sulphate of uranium and potassium — to a photographic plate enveloped in black paper. A darkening of the plate was observed, showing that this substance emitted rays capable of passing through matter opaque to ordinary light. It was soon found, however, that this property was not in any way the result of phosphorescence, for it was exhibited by all the compounds of uranium and the metal itself, even if these had been kept for a long time in a dark room.

The radiations emitted from uranium were found to be similar to X-rays in their penetrating powers. It was at first thought that they differed from X-rays in showing some evidence of reflection, refraction, and polarization, but this was found later to be incorrect.

Becquerel observed that the uranium rays, in addition to their photographic action, possessed, like X-rays, the important property of discharging an electrified body. This was later examined in detail by the writer,[2] who found that this discharging action could be explained on the assumption that the gas was ionized by the passage of the radiations through it. The ions were found to be identical with those produced by X-rays, and the ionization theory could consequently be directly applied to explain the various discharge phenomena produced by the rays from uranium. At the same time, it was clearly brought out that the rays from uranium consisted of two distinct kinds, called the α and β rays. The former were very easily absorbed in air and in thin sheets of foil, while the latter were of a far more penetrating type.

The intensity of the radiations emitted by uranium, whether examined by the photographic or electrical method, remains constant, or at any rate changes extremely slowly, for no appreciable alteration has been observed over a period of several years. The photographic and electrical effects exhibited by uranium

[1] Becquerel: Comptes rendus, cxxii, pp. 420, 501, 559, 689, 762, 1086 (1896).
[2] Rutherford: Phil. Mag., Jan., 1899.

are very feeble compared with those produced by the X-rays from the ordinary focus tube. An exposure to uranium salts of at least one day is required to produce any marked action on the plate.

The term "radioactivity" is now generally understood to signify the property shown by the class of substances, of which uranium, thorium, and radium are the best-known examples, of spontaneously emitting special types of radiations capable of acting on a photographic plate and of discharging an electrified body. The term "activity" is used to denote the intensity of the electrical or other effect of the rays from a substance compared with that shown by some standard substance. Uranium is usually chosen as this standard substance, in consequence of the constancy of its radiations, and the activity exhibited by other bodies is usually expressed in terms of the ratio of the electrical effect produced by the active matter under consideration compared with that of an equal weight either of uranium metal itself or of uranium oxide spread over an equal radiating area. For example, when the activity of radium is said to be about two millions, it is meant that the electrical effect due to it is two million times as great as the corresponding effect produced by an equal weight of uranium spread over an equal area.

Although the property possessed by uranium of spontaneously emitting energy in special forms without any apparent change in the matter itself could not fail to be regarded as a most remarkable phenomenon, yet the rate of emission of energy, judged by ordinary standards, is so feeble that it did not attract that active scientific attention which was afterwards excited by the discovery of radium; for this substance exhibited the properties of uranium to such a remarkable degree that it impressed both the lay and the scientific mind.

Shortly after the discovery of Becquerel, Mme. Curie[1] made a systematic examination of different substances for radioactivity, and found that the element thorium possessed a similar property to uranium and almost to the same degree. This fact was also

[1] Mme. Curie: Comptes rendus, cxxvi, p. 1101 (1898).

independently observed by Schmidt.[1] An examination was then made of the natural minerals which contain thorium and uranium, and here an unexpected result was observed. Some of these minerals were found to be several times more radioactive than pure uranium or thorium, and in all cases the uranium minerals showed four to five times the activity to be expected from the percentage of uranium present. Mme. Curie found that the radioactivity of uranium was an atomic property, *i. e.*, the activity observed depended only on the amount of the element uranium present, and was not affected by its combination with other substances. This being so, the large activity of the uranium minerals could only be accounted for by supposing that another unknown substance was present, which was far more active than uranium itself.

Relying on this hypothesis, Mme. Curie boldly proceeded to see if it were possible to separate chemically this unknown active substance from uranium minerals. By the courtesy of the Austrian Government, she was presented with a ton of uranium residues from the State Manufactory at Joachimsthal, Bohemia. In this locality there are extensive deposits of uraninite, commonly called pitchblende, which are mined for the uranium they contain. This pitchblende consists mainly of uranium, but also contains small quantities of a number of rare elements.

As a guide to the separation of the active substance, Mme. Curie employed a suitable electroscope to measure the ionization produced by the active body. After any chemical separation, the activities of the precipitate and of the filtrate evaporated to dryness were separately examined, and in this way it was possible to ascertain whether the active substance had been mainly precipitated or left behind in the filtrate.

The electric method was thus used as a rapid means of qualitative and quantitative analysis. Proceeding in this way, Mme. Curie found that not one but two very active substances were present in the uranium residues. The former of these, which was separated with bismuth, was called polonium,[2] in honor of

[1] Schmidt: Annal. d. Phys., lxv, p. 141 (1898).
[2] Mme. Curie: Comptes rendus, cxxvii, p. 175 (1898).

the country of her birth, and the latter, which was separated with barium, was called radium.[1] This latter name was a happy inspiration, for the activity of this substance in a pure state is at least two million times that of uranium. Mme. Curie then proceeded with the laborious work of separating the radium from the barium, and was finally successful in obtaining a small quantity of probably pure radium chloride. The atomic weight was found to be 225. The spectrum, first examined by Demarçay, was found to consist of a number of bright lines, and was analogous in many respects to the spectra of the alkaline earths.

In chemical properties radium is closely allied to barium, but can be completely separated from it by taking account of the differences in the solubility of the chlorides and bromides. On account of the small quantities of radium compounds available and of their great cost, no one has yet endeavored to obtain radium in a metallic state. Marckwald,[2] however, electrolyzed a radium solution with a mercury cathode, and concluded that the metal forms an amalgam with the mercury in the same way as barium. The small trace of the metal so obtained exhibited the characteristic radiating properties of the radium compounds.

There cannot be the slightest doubt that when radium is obtained in the metallic state it will be radioactive; for the property of radioactivity is atomic and not molecular. In addition, uranium and thorium as metals exhibit the activity to be expected from an examination of the activities of their compounds.

Radium exists in very small quantity in radioactive minerals. It will be seen later that the amount of radium in different minerals is always proportional to their content of uranium. The amount of radium per ton of uranium is about .35 gram, or less than one part in a million of the mineral. From a ton of Joachimsthal uraninite, which contains about 50 per cent of uranium, the theoretical yield of radium should be about .17 gram.

[1] M. and Mme. Curie and G. Bemont: Comptes rendus, cxxvii, p. 1215 (1898).

[2] Marckwald: Ber. d. d. chem. Ges., No. 1, p. 88, 1904.

In order to separate the radium from the barium mixed with it, Mme. Curie employed the method of fractional crystallization of the chlorides. Giesel[1] found that by use of the bromide instead of the chloride the separation of radium from barium was much facilitated. He states that six crystallizations suffice almost completely to remove the radium from the barium.

The discovery of radium gave a great impetus to the chemical examination of radioactive minerals in order to see if other radioactive substances were present. Debierne[2] succeeded in obtaining a new radioactive body called "actinium." Giesel[1] independently observed the presence of a new radiating body which he called the "emanating substance," and later "emanium," on account of the rapid emission from it of a short-lived radioactive emanation or gas. Recent work has shown that the substances separated by Debierne and Giesel are identical in radioactive properties and must contain the same element. Hofmann and Strauss[3] separated an active substance which was precipitated with lead and called by them "radiolead," while Marckwald[4] later obtained from pitchblende residues some extremely active matter which he named "radiotellurium," since it was initially separated with tellurium as an impurity.

None of these active bodies except radium have yet been obtained in a pure state. We shall see later that the active element present in the radiotellurium of Marckwald is almost certainly identical with that in the polonium of Mme. Curie; it will also be shown that the active elements present in radiolead and radiotellurium are in reality produced from the radium present in pitchblende, or, in other words, both are products of the transformation of the radium atom.

The possibility of using very active preparations of radium as a source of radiation led to a close examination of the nature of the rays emitted so freely from this substance. Giesel[5] observed in 1899 that the more penetrating rays, known as β rays,

[1] Giesel: Ber. d. d. chem. Ges., p. 3608, 1902; p. 342, 1903.

[2] Debierne: Comptes rendus, cxxix, p. 593 (1899); cxxx, p. 206 (1900).

[3] Hofmann and Strauss: Ber. d. d. chem. Ges., p. 3035, 1901.

[4] Marckwald: Ibid., p. 2662, 1903.

[5] Giesel: Annal. d. Phys., lxix, p. 834 (1899).

were deflected in a magnetic field in the same direction as the cathode rays, indicating that they consisted of negatively charged particles projected with great speed from the active substance.

This was substantiated by experiments of Becquerel,[1] who examined the deviation of a pencil of rays both in a magnetic and electric field. His results showed that the β particles had the same small mass as the particles of the cathode stream, whose corpuscular nature had previously been demonstrated by J. J. Thomson. The β particle was in fact identical with the electron set free by the electric discharge in a vacuum tube.

The β particles were projected from radium at different speeds, but their average velocity was much greater than that impressed on the electron in a vacuum tube, and in many cases approached closely to the velocity of light. This property of radium of emitting a stream of β particles at different speeds was later utilized by Kaufmann[2] to determine the variation of the mass of the β particle with velocity. J. J. Thomson had shown in 1887 that a charged body in motion possessed electrical mass in virtue of its motion. The theory of this action was later developed by Heaviside, Searle, Abraham, and others.

The moving charge acts as an electric current, and a magnetic field is generated round the body and moves with it. Magnetic energy is stored in the medium surrounding the charged body, which consequently behaves as if it had a greater apparent mass than when uncharged. This additional electric mass, according to the theory, should be constant for small speeds but should increase rapidly as the velocity of light is approached.

Kaufmann found from his experiments that the apparent mass of the electron did increase with speed, and that the increase was rapid as the velocity of light was approached. By comparing the theory with experiment, he concluded that the apparent mass of the β particle was entirely electrical in origin, and that there was no necessity to assume the presence of a material nucleus on which the charge was distributed.

This was a very important result, for it indirectly offered a

[1] Becquerel: Comptes rendus, cxxx, p. 809 (1900).
[2] Kaufmann: Physik. Zeit., iv, No. 1 b, p. 54 (1902).

possible explanation of the origin of mass, which has always been such an enigma to science. If a charge of electricity in motion exactly simulates the properties of mechanical mass, it is possible that the mass of matter in general may be electrical in origin, and may result from the movement of the electrons constituting the molecules of matter. Such a point of view, while most suggestive and important, cannot at present be considered more than a justifiable speculation.

Villard[1] in 1900 found that radium emitted in addition to α and β rays, a third type of radiation, now called the γ rays, which are of an extremely penetrating character. These rays are undeflected by a magnetic or electric field and appear to be a type of penetrating X-rays, which accompany the expulsion of the β particles from radium. The presence of these rays was also observed later in thorium, uranium, and actinium.

In the meantime, the importance of the α rays began to be more clearly recognized. These rays do not possess much power of penetrating matter, for they are stopped in their passage through a few centimetres of air and by a few thicknesses of metal foil. On the other hand, they produce far more ionization in the gas than the β and γ rays, and the greater proportion of the energy radiated from radioactive bodies is in the form of these rays. They were at first thought to be non-deflectable in a magnetic field, but in 1902 the writer[2] showed that they were deflected to a measurable extent in strong magnetic and electric fields. The direction of deflection was opposite to that of the β particles, showing that they carry a positive and not a negative charge of electricity.

From measurements of the amount of deflection of the rays both in a magnetic and electric field, it was found that the α particle from radium was projected with a velocity about $\frac{1}{10}$ that of light and had a mass about twice that of the hydrogen atom. The α rays from radium thus consist of a stream of atoms of matter projected with great velocity. We shall see later that there is some reason to believe that the α particle is an atom of

[1] Villard: Comptes rendus, cxxx, pp. 1010, 1178 (1900).

[2] Rutherford: Phil. Mag., Feb., 1903; Physik. Zeit., iv, p. 235 (1902).

helium. The main radiations from radium are thus corpuscular in character, and consist of streams of positively and negatively charged particles.

In 1903, Sir William Crookes,[1] and Elster and Geitel,[2] independently observed a very interesting property of the α rays. The α rays from radium or other strongly active substance produce phosphorescence on a screen of crystalline zinc sulphide (Sidot's blende). On examination of the screen with a lens, the luminosity is found to be not uniform, but to consist of a number of bright points of light, which follow one another in irregular but rapid succession. These "scintillations" are a result, probably indirect, of the bombardment of the screen by the massive α particles, but the exact explanation of this striking phenomenon is still unsettled.

In the meantime the complexity of the processes occurring in thorium and radium became more evident. The writer[3] in 1900 showed that thorium, in addition to the expulsion of α and β particles, continuously emits a radioactive emanation or gas. Both radium and actinium also exhibit a similar property. These emanations consist of gaseous radioactive matter, the radiating power of which rapidly dies away. The emanations of thorium, radium, and actinium can readily be distinguished from one another by the rate at which they lose their activity. The emanations of both actinium and thorium are very short lived, the former losing half of its activity in 3.9 seconds and the latter in 54 seconds. On the other hand, the emanation from radium is far more persistent, and requires a lapse of about four days to reduce the activity to half value.

About the same time another remarkable action of radium and thorium was disclosed. M. and Mme. Curie[4] found that all bodies placed in the neighborhood of radium salts became temporarily active. A similar property was independently observed by the writer for thorium.[5] This property of radium and

[1] Crookes: Proc. Roy. Soc., lxxxi, p. 405 (1903).

[2] Elster and Geitel: Physik. Zeit., No. 15, p. 437, 1903.

[3] Rutherford: Phil. Mag., Jan. and Feb., 1900.

[4] M. and Mme. Curie: Comptes rendus, cxxix, p. 714 (1899).

[5] Rutherford: Phil. Mag., Jan. and Feb., 1900.

thorium of "exciting" or "inducing" activity in substances placed near them is directly due to the emanations from these bodies. The emanation is an unstable substance and is transformed into a non-gaseous type of matter which is deposited on the surface of all bodies in its neighborhood.

Another striking property of radium was observed by P. Curie and Laborde in 1903.[1] A radium compound continuously emits heat at a rate sufficient to melt more than its own weight of ice per hour. In consequence of this a mass of radium always keeps itself at a higher temperature than the surrounding air. This rapid emission of heat by radium is directly connected with its radioactive properties, and will be shown later to result mainly from the bombardment of the radium by the α particles projected from its own mass.

From the above brief review of the more important properties exhibited by the radioactive bodies, it will be seen that the processes taking place in a mass of radioactive matter are very complicated. In a compound of radium, for example, there occurs a rapid expulsion of α and β particles accompanied with the generation of γ rays, a rapid emission of heat, the continuous production of an emanation or gas, and the formation of an active deposit which gives rise to "excited" activity.

A great advance in the clear understanding of the connection between these various processes resulted from the discovery by Rutherford and Soddy[2] that a very active substance, called thorium X, could be separated from thorium by a simple chemical operation. This thorium X was found to lose its activity in time, while the thorium freed from ThX spontaneously produced a new supply of ThX. In a mass of thorium in radioactive equilibrium, the two processes of growth and change of ThX proceed simultaneously, and the amount of ThX present reaches a constant value when its rate of production from the thorium balances its own rate of change. It was found that the thorium "emanation" was directly produced by ThX, while

[1] P. Curie and Laborde: Comptes rendus, cxxxvi, p. 673 (1903).

[2] Rutherford and Soddy: Phil. Mag., Sept. and Nov., 1902; Trans. Chem. Soc., lxxxi, pp. 321 and 837 (1902).

the emanation in turn gave rise to the active deposit which causes the phenomenon of excited activity.

Now it has been pointed out that the radioactive property is atomic, and consequently must result from a process occurring in the atom and not in the molecule. In order to explain the results observed, Rutherford and Soddy advanced a theory, known as the "disintegration theory." It is supposed that the atoms of the radioactive bodies are unstable, and that a certain fixed proportion of them become unstable every second and break up with explosive violence, accompanied in general by the expulsion of an α or β particle or both together. The residue of the atom, in consequence of the loss of an α particle, is lighter than before, and becomes the atom of a new substance quite distinct in chemical and physical properties from its parent. In thorium, for example, it is supposed that the atom of ThX consists of the thorium atom minus an α particle. Thorium X is unstable and breaks up at a definite rate with the expulsion of another α particle. The residue of the atom of ThX in turn becomes the atom of the emanation, and this in turn breaks up through a further succession of stages.

The theory was found to account in a satisfactory way for the processes occurring not only in thorium but in all the radioactive bodies. On this view, the radioactive substances are undergoing spontaneous transformation with the appearance of a number of new kinds of matter which are unstable and have a limited life. The radiations accompany the transformations and are produced as a result of an explosive disturbance within the atom.

The long continued emission of energy from the radioactive bodies does not on this view present any fundamental difficulty and is in accordance with the principle of the conservation of energy. The matter loses in atomic energy at each stage of the transformation, and the energy radiated is derived from the internal energy resident in the atoms themselves. The atom is supposed to consist of a number of charged parts in rapid oscillatory or orbital motion and consequently contains a great store of energy. Part of this energy is kinetic and part potential, resulting from the condensation of the electrical charges within the minute volume of the atom. This latent energy of the atom does not ordinarily manifest itself, since the chemical and physical forces at our disposal do not allow us to break up the atom. Part of this energy is, however, released in radioactive changes where the atom itself suffers disruption with the expulsion of one of its charged parts with great velocity.

This theory has proved of the greatest service in correlating the various phenomena shown by the active substances. In many cases, it offers a quantitative as well as a qualitative explanation of the experimental facts, and has proved of great value in suggesting new lines of attack.

In addition to its aid in tracing the succession of transformations which occur in the radioelements, this theory has been instrumental in showing that radium is produced from uranium and that the active constituents in radiolead and radiotellurium result from the transformation of radium.

The application of this theory to the unravelling of the complicated series of transformations in radium, thorium, and actinium will form the main subject matter of this treatise.

The disintegration theory received a strong measure of support from the remarkable observation of Ramsay and Soddy[1] that the rare gas helium was produced from the radium emanation. This in itself supplied unequivocal evidence that there was an actual transformation of matter taking place in radium, one of the products of which was the inactive gas helium.

It will be seen later that the weight of evidence points to the conclusion that the α particle from radium is an atom of helium. On this view helium arises during the transformation of each product which emits α rays. Such a conclusion, apart from other evidence, is also supported by the recent observation of Debierne that helium is produced from actinium as well as from radium.

In the above review we have traced the main line of advance of knowledge in the field of radioactivity, but there has been a rapid and important advance in another direction.

[1] Ramsay and Soddy: Proc. Roy. Soc., lxxii, p. 204 (1903); lxxiii, p. 341 (1904).

Elster and Geitel[1] showed in 1901 that radioactive matter existed in the atmosphere. Later work has shown that the radioactivity of the atmosphere is mainly due to the presence of the radium emanation which diffuses into the atmosphere from the earth. Elster and Geitel and others have made an extensive examination of the radioactivity of soils, and of well and spring waters, and have shown that there is a very wide diffusion of small quantities of radioactive matter throughout the crust of the earth and in the atmosphere. A number of investigators have entered this new field of inquiry, and a large amount of valuable data has already been accumulated.

While the radioactive elements exhibit the quality of radioactivity in a very marked degree, there is an increasing body of evidence that ordinary matter also possesses this property to a very minute extent, and that the activity observed cannot be ascribed to the presence of traces of the known radioelements. This detection of the radioactivity of ordinary matter has been made possible by the extraordinary delicacy of the electrical test for the presence of radiations which are able to ionize a gas.

When it is remembered that the initial discovery of the radioactivity of uranium was made in 1896, and that the first evidence of the presence of radium was obtained in 1898, it will be seen how rapid has been the advance in our knowledge of this complicated subject. A very large mass of experimental facts has now been accumulated, and their connection with one another has been made clear by the adoption of a simple theory. The rapidity of this advance has seldom, if ever, been equalled in the history of science, and it is of interest to examine the influences that have made it possible.

It is not due to the number of workers in the field, for until the last year or two the subject has been represented by comparatively few investigators. The main reason for the rapidity of the advance lies in the remarkably opportune time at which the new field was opened up, and the influence upon it of the rapid extension of our knowledge of the passage of electricity through gases.

[1] Elster and Geitel: Physik. Zeit., ii, p. 590 (1901).

In this connection it is of interest to note that the discovery of the radioactive property of uranium might accidentally have been made a century ago, for all that was required was the exposure of a uranium compound on the charged plate of a gold-leaf electroscope. Indications of the existence of the element uranium were given by Klaproth in 1789, and the discharging property of this substance could not fail to have been noted if it had been placed near a charged electroscope. It would not have been difficult to deduce that the uranium gave out a type of radiation capable of passing through metals opaque to ordinary light. The advance would probably have ended there, for the knowledge at that time of the connection between electricity and matter was far too meagre for an isolated property of this kind to have attracted much attention.

It is not necessary, however, to go so far back to illustrate the great influence that the electrical discoveries in the allied field of discharge of electricity through gases have had on the rapid development of radioactivity. If the discovery had been made even a decade earlier, the advance must necessarily have been much slower and more cautious. At that time the possibility of the existence of radiations capable of penetrating matter opaque to light had not even been considered, and the true nature of the cathode rays was still a matter for conjecture. The character of the radiations from radioactive matter, as we know them to-day, could only have been deduced after a long and laborious series of researches, for not only would the experimenter have had no guidance from analogy, but he must of necessity have developed the methods of attack *de novo* under difficult conditions. It would have been necessary, also, to have examined in detail the nature of the discharging action of the rays, for on this is based the most important method of measurement in radioactivity.

Let us now examine the conditions that existed during the actual development of the subject. We have seen that the mechanism of the conduction of electricity through gases had been developed primarily by a study of the conductivity of gases exposed to X-rays, and of the discharge of electricity in a vacuum tube. The knowledge thus obtained was directly applied to the corresponding ionization produced by the radiations from active matter and served as a foundation for the electric method of measurement which has been utilized as a rapid quantitative means of radioactive analysis. When the β rays of radium were found to be deflected in a magnetic field in the same way as the cathode rays, it was only necessary, in order to prove their identity, to adopt the methods which had been familiar to science for several years. In a similar way, the behavior of the non-deflectable γ rays was directly compared with the known properties of X-rays, while the α rays were found to be analogous in some respects to the canal rays of Goldstein, which had been shown previously by Wien to be deflected by a magnetic and electric field.

The influence of the ionization theory on the development of radioactivity has been equally marked in other directions. The determination of the charge carried by an ion has been of the greatest utility in determining the order of magnitude of the processes occurring in radioactive matter. These data have been of great value in determining the number of α and β particles emitted by radium and in suggesting the probable amount of emanation and of helium liberated from it. Such calculations have enabled us to fix with some certainty the rate at which radium and the other radioactive bodies suffer disintegration, and also to determine beforehand the magnitude of many physical and chemical quantities, and have thus indirectly suggested the methods of attack necessary to solve the various problems which have arisen.

The fortunate combination of events in the history of radioactivity is strikingly illustrated by the discovery that helium is evolved by the radium emanation. This rare gas has a dramatic history, for its presence was first observed in the sun by Janssen and Lockyer in 1868; but it was not until 1895 that its presence in the rare mineral clevite was observed by Ramsay. An examination of its physical and chemical properties had hardly been completed when Ramsay and Soddy, guided by the disintegration theory, made an examination of the gases liberated from radium, and discovered that helium was a product of the transformation of radium. If helium had not a short time before been found in radioactive minerals, it is safe to say that this most striking property of radium of producing helium would have long remained hidden.

While the ionization theory has played a prominent part in the extension of our knowledge in radioactivity, the benefits have not been altogether one-sided, for the results obtained from a study of radioactivity have greatly assisted in extending and confirming the ionization theory. It has placed in the hands of the experimenter a constant and powerful source of ionizing radiation in place of a variable source like X-rays, and this has been of great service in obtaining accurate data. In addition, the results of Kaufmann on the variation of the mass of the β particle from radium with its velocity, have been an important factor in confirming and extending our conception of electrons.

Examples of this kind could readily be multiplied, but sufficient have been given to illustrate the close connection that has existed and still exists between these two distinct lines of investigation, and the influence which each has exerted upon the development of the other.

Radiations from Active Bodies

A brief summary will now be given of the chief properties and nature of the α, β, and γ rays emitted from radioactive bodies. All three types of rays possess in common the properties of acting on a photographic plate, of exciting phosphorescence in certain substances, and of discharging electrified bodies. The three types of rays can be distinguished from one another by their difference in penetrating power and by the effect upon them of a magnetic or electric field. The α rays are completely stopped by a layer of aluminium about .05 mm. in thickness; the greater part of the β rays by 5 mms. of aluminium, while a thickness of at least 50 cms. of aluminium would be required to absorb most of the γ rays. The relative penetrating power of the three types of rays is thus about in the ratio 1 : 100 : 10000. It must be borne in mind, however, that this is an average value, for each type of radiation is complex and consists of rays unequally absorbed by matter.

The α rays consist of positively charged particles projected with a velocity of about twenty thousand miles per second. The apparent mass of the α particle is about twice that of the hydrogen atom. Although the magnetic deflection of these rays has only so far been observed in the case of active substances like radium and polonium, there can be little doubt that the α rays from the other radioactive bodies are of a very similar nature.

The β rays consist of negatively charged particles projected with great velocities. Their apparent mass is about $\frac{1}{1000}$ of that of the hydrogen atom, and they are identical in all respects except velocity with the cathode-ray particles set free in a vacuum tube.

In the case of radium, the β particles are projected with a wide range of velocity, the maximum approaching very closely to the velocity of light. β rays are also given out by uranium, thorium, and actinium.

The γ rays are not deflected by a magnetic or electric field, and closely resemble in general properties the very penetrating X-rays produced in a hard vacuum tube. According to present views, the γ rays must thus be supposed to be a type of wave motion in the ether, consisting probably of pulses set up as a result of the expulsion of β particles. Only those active substances which emit β rays give rise to γ rays. The γ rays are emitted from uranium, thorium, radium, and actinium, but the rays from uranium and actinium have not such a marked power of penetration as the rays from thorium and radium.

Each of these three primary types of rays falling upon matter gives rise to secondary rays. In the case of the α rays the secondary radiation consists of negatively charged particles (electrons) projected at velocities comparatively small compared with those of the β particles themselves. The secondary rays arising from the β and γ rays consist in part of electrons projected with considerable velocity. These secondary rays in turn produce tertiary rays, and so on.

If a strong magnetic field is applied at right angles to a pencil of α, β, and γ rays, the three types of rays are separated from each other. This is shown in Fig. 1, where the magnetic field is acting downwards, perpendicular to the plane of the paper. The β rays are bent to the right, the α rays to the left, while the γ rays are unaffected. The β rays consist of particles having unequal velocities and consequently traversing circular orbits of different radii of curvature. The magnetic deflection of the α rays compared with that of the β rays is much exaggerated in the figure.

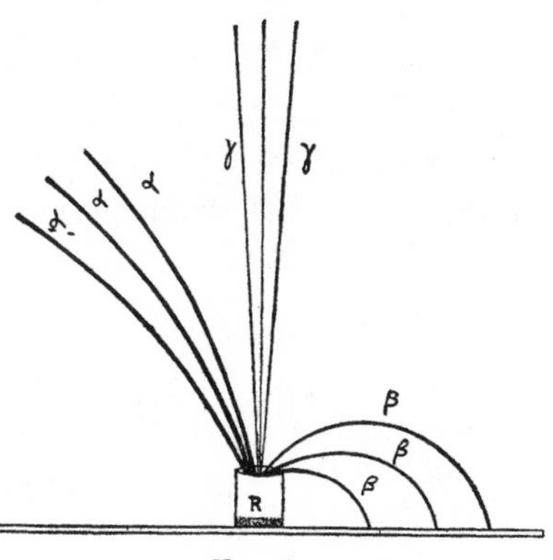

FIG. 1.

Separation of radium rays by the action of a magnetic field.

The relative mass, velocity, and kinetic energy of the average α and β particles are shown in Fig. 2, where the volume of a sphere represents mass and energy, and the length of line represents velocity.

It will be seen from this illustration that although the average β particle has a much higher velocity than the average α particle, its energy of motion on account of its relatively small mass is much less than that of the average β particle. This result is in accordance with the observed result that the ionization and heating effect produced by the α particle are much greater than for the β particle.

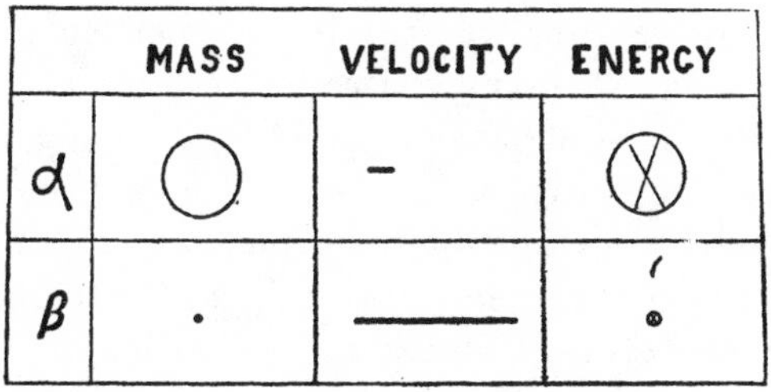

	MASS	VELOCITY	ENERGY
α			
β			

FIG. 2.

The writer has recently shown that one gram of radium in

radioactive equilibrium emits about 7×10^{10} β particles and about 2.5×10^{11} α particles per second. Four α particles are thus expelled from radium for each β particle.

Radioactive Substances

Below is given a list of the radioactive substances which have so far been separated. The nature of the radiations, and the presence or absence of an emanation is also noted. The "period" of the emanation denotes the time required for its activity to fall to half value.

Uranium: α, β, and γ rays, but no emanation.
Thorium: α, β, and γ rays; an emanation, period 54 seconds.
Radium: α, β, and γ rays; an emanation, period 3.8 days.
Actinium / Emanium: α, β, and γ rays; an emanation, period 3.9 seconds.
Polonium / Radiotellurium: only α rays; no emanation.
Radiolead (some time after preparation): α, β, and γ rays, but no emanation.

These substances, with the exception of polonium, continue to radiate for long periods of time. In addition, there are a number of radioactive products arising from each radioelement which have a comparatively short radioactive life. These products are intrinsically as important as the more permanently active substances, and have an equal right to be called elements. On account of the rapidity of their transformation, they exist in extremely small quantity in pitchblende, and are never likely to be obtained in sufficient quantity to be examined by ordinary chemical methods. Polonium and radiotellurium, which contain the same radioactive constituent, differ from the other substances in emitting only α rays. In regard to their length of life, they occupy an intermediate position between the rapidly transformed products like the emanation, and a very slowly changing substance like radium. The activity of radiotellurium falls to half value in about 140 days, while the corresponding time for radium is about 1300 years.

With the exception of uranium, thorium, and radium, none of these substances have been sufficiently purified to determine their atomic weight or spectrum. It seems likely, however, that actinium will prove to be an element at least as active as radium. It will also be shown later that radiotellurium and radiolead in a pure state should be much more active, weight for weight, than radium.

The activity of a given substance which emits α rays depends on the number of α particles shot out per second, and this, for equal weights, is inversely proportional to the "period" of that substance. For example, the actinium emanation whose period is 3.9 seconds must be, weight for weight, at least one thousand million times as active as radium. It is on account of their enormous activity and consequent rapidity of transformation that such substances can never be obtained in sufficient quantity for chemical analysis. It is only the more slowly changing substances like radium, radiolead, and radiotellurium that collect in sufficient quantity in pitchblende to be chemically isolated in appreciable quantity.

It will also be shown later that the radiations emitted from uranium, radium, thorium, and actinium arise only in part from the primary active substance itself. The β and γ rays in all cases are emitted from the products of the transformation of these elements. These are mixed with the parent substance and add their radiations to it.

Methods of Measurements

There are three general properties of the rays from radioactive substances which have been utilized for the purpose of measurements, depending on (1) the action of the rays on a photographic plate, (2) the phosphorescence excited in certain crystalline substances, (3) the ionization produced by the rays in a gas. Of these the phosphorescent method is limited to substances like radium, actinium, and polonium which emit very intense radiations. The α, β, and γ rays all produce a marked luminosity in the platinocyanides and in the mineral willemite (zinc silicate). The mineral kunzite responds mainly to the β

and γ rays, while Sidot's blende (crystalline zinc sulphide) responds mainly to the α rays. Besides these there are a large number of substances in which a more or less feeble luminosity is excited by the rays. The property of the α rays of producing scintillations on a screen covered with zinc sulphide is especially interesting, and it has been found possible by this method to detect the α rays emitted by feebly active substances like uranium, thorium, and pitchblende. Screens of zinc sulphide have been used as an optical method for demonstrating the presence of the emanations from radium and actinium. Speaking generally, the phosphorescent method, while very interesting as an optical means for examining the rays, is very limited in its application and is only roughly quantitative.

The photographic method proved of great service in the early development of radioactivity, but has gradually been displaced by the electric method as quantitative determinations have become more and more necessary. It has proved of special utility in examination of the curvature of the path of the rays in magnetic and electric fields. On the other hand, it does not readily lend itself to quantitative comparisons and is very limited in its application. In the case of feebly active substances like uranium and thorium, long exposures are necessary to produce much photographic effect. It cannot be utilized to follow the rapid changes of activity which are exhibited by many radioactive products, and is not sufficiently sensitive to detect the presence of rays which are readily observed by the electric method.

The development of the subject of radioactivity has largely depended on the electric method of measurement, which is universally applicable, and far transcends in delicacy either of the other two methods. It readily lends itself to rapid quantitative measurements, and can be applied to all the types of radiation which possess the ionizing property.

This method is, as we have seen, based on the property of the α, β, and γ rays of producing charged carriers or ions in the volume of the gas traversed by the radiations. Suppose that a layer of radioactive substance — uranium, for example — is placed on the lower of two insulated parallel plates, A and B. (Fig. 3).

The gas between the plates is ionized at a constant rate by the radiations, and there results a distribution of positive and negative ions in the volume of air. If no electric field is acting, the number of ions does not increase indefinitely, but soon reaches a maximum, when the rate of production of fresh ions by the radiations exactly compensates for the decrease in the number due to the recombination of the positive and negative ions. This latter effect will obviously tend to take place when the positive and negative ions in the course of their movement come within the sphere of one another's attraction. Suppose now that the plate A is kept charged to a constant potential V, and

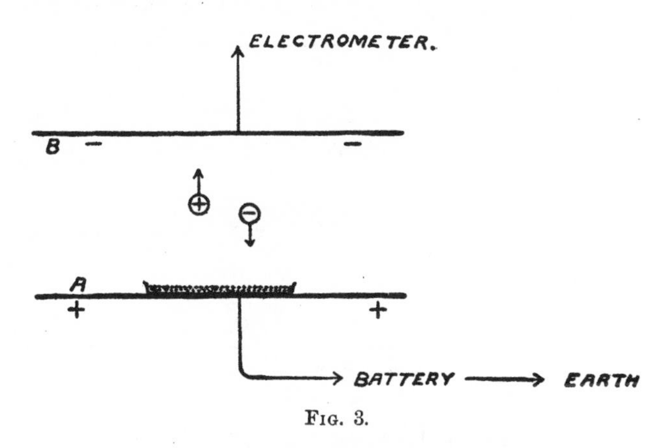

Fig. 3.

that the rate at which B, initially at zero potential, gains an electric charge is determined by a suitable measuring instrument, for example, a quadrant electrometer.

Under the influence of the electric field, the positive ions travel to the negative plate and the negative ions to the positive. There is consequently a current through the gas, and the plate B and its connections acquire a positive charge. The rate at which the plate B rises in potential is a relative measure of the current through the gas. When V has a small value, the current is small, but gradually increases with rise of V, until a stage is reached where the current increases very slightly for a large increment of the value of V. The relation between the current and the applied voltage is seen in Fig. 4. The shape of this curve receives a simple explanation on the ionization theory. The ions move with a velocity proportional to the strength of the electric field. In a weak field there is thus a slow movement of the positive and negative ions past one another. A large proportion of the ions have time to recombine before they reach the electrodes, and the current observed through the gas is consequently small. As the voltage increases, the velocity of the ions increases, and there is less time for recombination. Finally, in a strong field practically all the

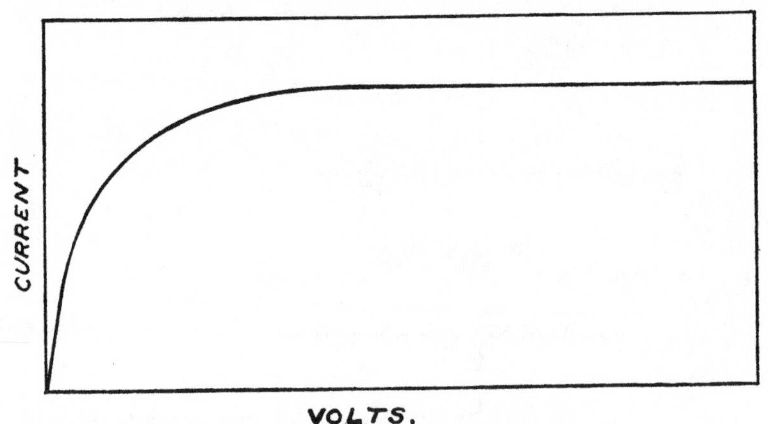

Fig. 4.
Typical saturation curve for an ionized gas.

ions are swept to the electrodes before any appreciable recombination can occur. The maximum or "saturation" current through the gas is then a measure of the charge carried by the ions produced per second by the radiation, *i. e.*, it is a measure of the total rate of production of ions.

The term "saturation," which was applied initially from the resemblance of the current-voltage curve to the magnetization curve for iron, is not very suitable, but has come into use as a convenient though inaccurate method of expressing an experimental fact.

Other conditions being the same, the voltage required to produce saturation increases with the intensity of the ionization, *i. e.*, with increase in activity of the substance under examination. Increase of the distance between the plates lowers the value of the electric field and increases the distance over which the ions move. Both of these conditions tend to increase the voltage to be applied in order to give saturation.

It is found experimentally that for parallel plates not more than 3 or 4 cms. apart, 300 volts is sufficient to produce approximate saturation, using substances whose activities are not greater than 1000 times that of uranium. For intensely active substances like radium, in order to produce saturation at all, the plates must be close together and a high voltage applied.

The essential condition for quantitative comparisons by the electric method depends on the measurement of the saturation current, for this is a measure of the total number of ions produced per second in the volume of gas under consideration.

The electric method can be used with accuracy to compare the relative activity of substances which emit identically the same rays, but differ only in the intensity of the radiations. It serves, for example, to determine accurately the rate at which simple products like the emanations lose their activities.

Unless other factors are taken into consideration, the electric method cannot be directly used to compare the relative intensity of different types of radiation. For example, the relative saturation currents produced by the α and β rays, emitted from a thick layer of uranium, under the conditions shown in Fig. 3, cannot be used as a direct comparison of the intensity of the two types of radiations, for on account of their difference in penetrating power, a much smaller fraction of the total energy of the issuing β rays is absorbed in producing ions between the plates than in the case of the more easily absorbed α rays. Before such comparisons of ionization currents can be used to measure the relative amounts of energy of the two types of rays, the relative penetrating and ionizing powers of the types of radiation must be accurately known. The main province of the electric method, however, lies in the determination of the variations in the activity of a body which emits rays all of one kind, and in this it has proved of great value, and has yielded results of considerable accuracy.

A variety of methods have been employed to measure the ionization currents produced by the radiations. If a very active substance is under examination, a sensitive galvanometer may be used to measure the saturation current. With slight modifications, the gold-leaf electroscope has proved an accurate and reliable means of measurement, and has played a prominent part in the development of radioactivity. Various types of the instrument have been employed. A simple form which I have found very convenient in comparisons of activities is shown in Fig. 5. The active material is placed on the lower plate, A, which is mounted on a slide so that it can be easily moved out to place the radioactive material in position. The upper plate, B, placed about 3 cms. above A, is connected with a rod, R, which is rigidly supported by a cross rod, TT, on two insulating sulphur rods, SS. The aluminium or gold leaf is connected with the upper part of the rod R. The rod C, when connected with a suitable voltage, serves to charge the electroscope system.

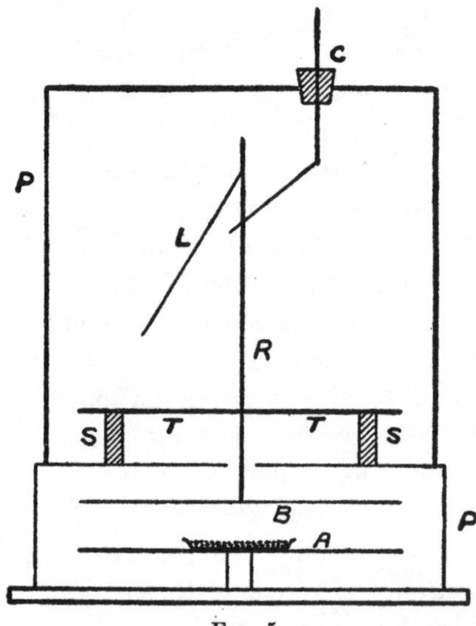

Fig. 5.
Simple electroscope for comparing α-ray activities.

The movement of the gold leaf is observed through glass or mica windows by means of a low-power microscope provided with a micrometer scale in the eyepiece. The lower plate, A, and the external case, PP, are connected to earth.

By suitably adjusting the length of the gold leaf and the position of the boundaries, the time taken for the gold leaf to pass over a definite number of divisions of the scale in the eye-piece may be made nearly constant over a considerable range. The radioactive material, placed in a metal or other conducting vessel, is put in position. The electroscope is then charged and the time taken for the gold leaf to pass over a fixed part of the scale is observed. This must be corrected for the natural leak of the instrument, which is determined before the introduction of the active material. This natural leak may be due in part to a slight leakage over the sulphur supports, or more generally to a weak activity of the walls of the electroscope. All substances are slightly active, and this activity is often increased by radioactive contamination from the presence of radium and other emanations. Two or three hundred volts is sufficient to charge the electroscope, and this insures saturation over the greater part of the range provided that the active material does not cause the electroscope to lose its charge in less than two or three minutes.

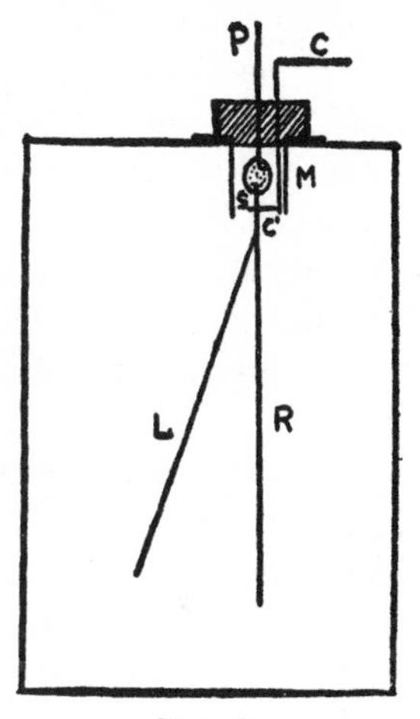

FIG. 6.

Electroscope for comparing β and γ ray activities and for measurement of very weak activities.

In this way measurements can be made with rapidity and certainty. An accuracy of one per cent can readily be obtained, and with care the precision of measurement may be made still greater. The great advantages of this type of instrument are its simplicity, portability, and comparative ease of construction. Such an instrument, if standardized by a constant source of radiation like uranium, is very suitable for determining the variation of activity of substances, which change very slowly with time.

A modification of this electroscope, first used by C. T. R. Wilson, can be utilized to measure extraordinarily minute ionization currents. The construction of the apparatus is seen in Fig. 6.

A clean metal vessel, preferably of brass, of about one litre capacity, has a gold leaf, L, attached to a rod, R, insulated by a sulphur or amber bead, S, inside the vessel. This is charged by a movable rod, C, or by a magnetic device. After charging, the upper rod, P, is connected to the case of the instrument and to earth. In special cases, if extremely minute currents are to be measured, the rod P is kept connected with a source of potential slightly greater than the potential of the electroscope system. This insures that there is no leak of the charge across the sulphur support.

The movement of the gold leaf is observed as before with a microscope having a micrometer eyepiece. The great advantage of this instrument lies in the fact that the apparatus can be hermetically closed. The rate of leak observed must then be entirely due to the ionization in the interior of the vessel and be independent of external electrostatic disturbances.

An instrument of this kind is very useful for comparing β and γ ray activities. For the former, the base of the electroscope is removed, and replaced by a sheet of aluminium, about .1 mm. in thickness, which completely stops the α rays, but allows the β rays to pass through with little absorption. For measurements of the γ rays, the vessel is placed on a lead plate about 5 mms. thick, and the active material placed beneath. The β rays are completely absorbed by this thickness of lead, and the ionization of the vessel is then due to the more penetrating γ rays.

The most convenient general method of measurement depends on the use of a quadrant electrometer. A very convenient and useful type of electrometer for radioactive and other work has been designed by Dolezalek.[1] The general construction of the instrument is seen in Fig. 7.

The four quadrants are mounted on amber or sulphur supports. A very light needle, N, is made out of silvered paper and is suspended by a fine quartz fibre or phosphor-bronze strip. The needle is charged to a potential of 100 to 300 volts. If a quartz suspension is used, this is done by lightly touching the metallic support of the needle with a wire connected with the

[1] Dolezalek: Instrumentenkunde, p. 345, 1901.

source of potential. It is often more convenient to use a fine phosphor-bronze suspension. The needle may then be directly connected with one terminal of a battery the other pole of which is earthed, and its potential kept constant. By the use of a fine quartz suspension, the sensibility of the instrument, *i. e.*, the number of millimetre divisions, passed over by the spot of light

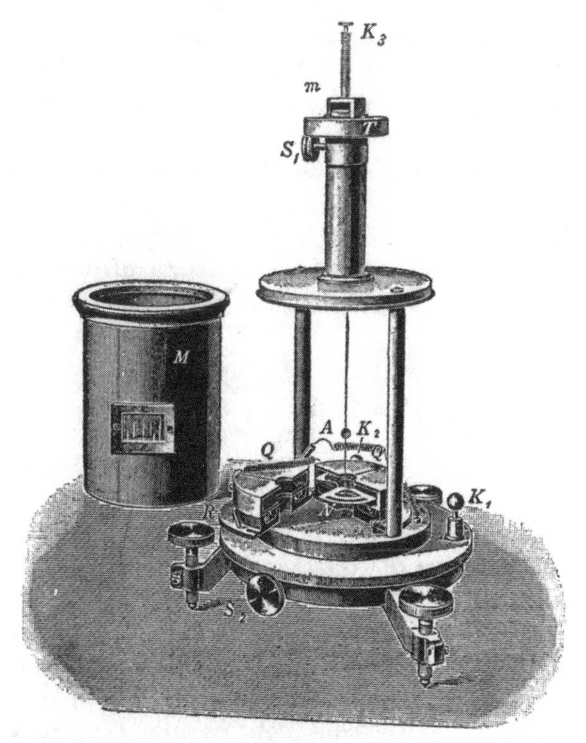

FIG. 7.
Dolezalek electrometer.

on the scale for an application of a difference in potential of one volt between the quadrants, may be made very great. A sensibility of 10,000 millimetre divisions per volt is not uncommon. Unless, however, a very small current is to be measured, it is not advisable to work with a sensibility greater than 1000 divisions per volt, and 200 divisions is often sufficient.

The quadrant electrometer is essentially an instrument for measuring the potential of the conductor with which it is connected, but is indirectly used in radioactivity to measure ionization currents. The capacity of the electrometer and its connections remains sensibly constant with the movement of the needle, and the rate at which the spot of light moves over the scale is a measure of the rate of rise of potential of the electrometer system. This serves as a measure of the ionization current between the electrodes of the testing vessel.

The general arrangement for measurement is shown in Fig. 8. The active material is placed on the lower of two parallel insulated plates, A and B. The plate A is connected with one

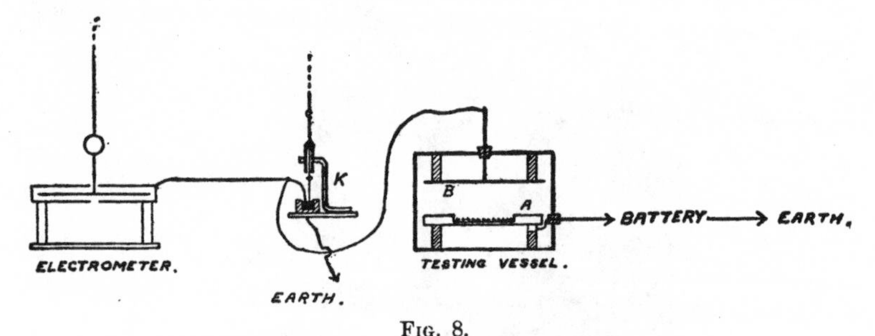

Fig. 8.

Method of use of electrometer for comparing activities.

terminal of a battery of suitable potential, and the plate B is connected through a key, K, with one pair of quadrants of the electrometer. When not in use the key connects the quadrants and its connections to earth. When a measurement is required, the earth connection is quietly but quickly broken by means of some mechanical or electro-magnetic device. The plate B and its connections rise in potential, and this is indicated by the movement of the spot of light over the scale. The time taken for the spot of light to move over a definite distance of the scale is then observed, and the number of divisions passed over per second serves as a comparative measure of the ionization current through the gas.

When the movement of the spot of light becomes too rapid for accurate observation, an additional capacity in the form of an air or mica condenser is added to the electrometer system and the rate of movement is reduced to that required.

Proceeding in this way, activities can readily be compared over a very wide range. The magnitude of the current that can be measured is only limited by the capacity of the condensers available and the voltage of the battery, which must be sufficient to produce saturation in the testing vessel.

This use of electroscopes and electrometers to compare activities thus depends on the rate of angular movement of the moving system. By a suitable arrangement, an electrometer may be used as a direct reading instrument for measuring current in the same way as a galvanometer.

Suppose that the electrometer system is connected to earth through a very high resistance, which obeys Ohm's law. When the earth connection of the quadrants is broken, the plate B (Fig. 8) rises in potential until the supply of electricity to B exactly compensates for the loss of electricity by discharge through the high resistance. Since the deflection of an electrometer needle is proportional to the voltage applied, the spot of light will thus move from rest to a steady position, and this deflection is proportional to the ionizing current in the testing vessel.

For measurements of this character, the resistance used must be of the order 10^{11} ohms. The main drawback of the method is the difficulty of obtaining suitable resistances of this character which are at the same time free from polarization and obey Ohm's law. The principle of this method has been employed in experiments by Dr. Bronson[1] in the laboratory of the writer.

Such an arrangement is especially suitable for following with accuracy rapid variations of activity. The deflection is independent of the capacity of the electrometer and connections, and measurements can be made quickly and accurately over a wide range.

Some of the types of testing vessels suitable for comparisons of activity by the electrometer method are shown in Figs. 9 and 10.

[1] Bronson: Amer. Journ. Science, July, 1905; Phil. Mag., Jan., 1906.

The active material is placed on the lower of two parallel plates, A and B, in a closed vessel (Fig. 9). The insulated plate, B, is attached by ebonite rods to the case of the apparatus

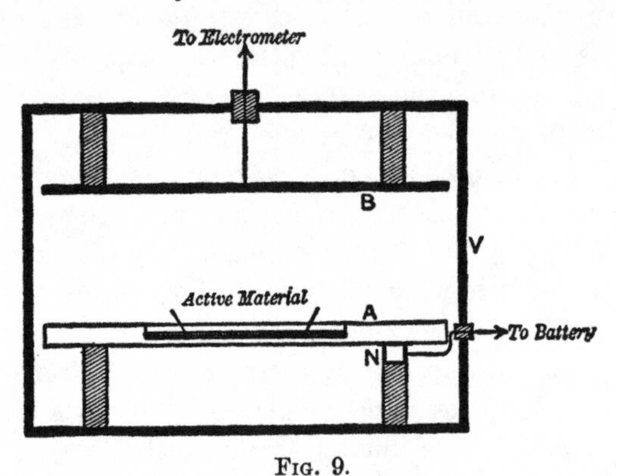

FIG. 9.

Parallel plate testing vessel.

which is connected with earth, so that there can be no direct conduction leak from the battery to B.

In Fig. 10 is shown a cylindrical testing vessel, B, suitable for comparison of the activities obtained on wires or rods. The

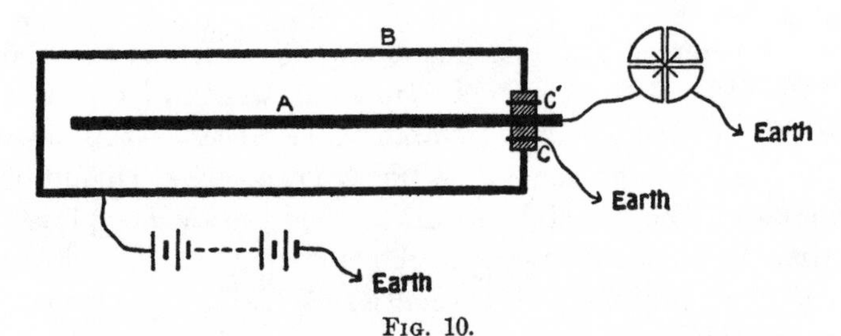

FIG. 10.

Cylindrical testing vessel with guard ring for comparing activities on wires or rods.

inner active electrode, A, passes through an ebonite cork. The conduction leak across the ebonite is avoided by use of the guard ring principle, a metal cylinder, CC′, connected with earth dividing the ebonite cork into two parts. In this case, the ebonite has only to be insulated sufficiently for the small rise of potential required to cause a suitable deflection of the electrometer needle. The adoption of the guard ring principle is advisable in all cases in order to get rid of possible conduction currents across the surface of the insulator.

An apparatus of this kind is very suitable for determining the decay curves of the excited activity imparted to cylindrical electrodes, and for determinations of the decay of activity of emanations which are introduced into the cylinder.

The electric method is an extraordinarily delicate means of detecting the presence of minute quantities of radioactive matter. This can be readily illustrated by a simple experiment. A milligram of radium bromide is taken and dissolved in 100 c.c. of water. After thorough mixing, 1 c.c. of this solution is taken and added to 99 c.c. of water. One c.c. of this last solution thus contains 10^{-7} gram of radium bromide. If this is evaporated on a metal vessel, the activity possessed by this minute quantity of radium suffices to cause an extremely rapid discharge when brought near the cap of an electroscope, such as is shown in Fig. 11. If the radium covered plate is placed on the cap of the electroscope, it is impossible for the leaves to retain their charge more than a few seconds.

Using an electroscope of small natural leak, the presence of 10^{-11} gram of radium can easily be detected by the increased rate of movement imparted to the gold leaf.

Extraordinarily minute currents can be measured with accuracy in an electroscope of the type shown in Fig. 6. For example, Cooke observed that in a well cleaned brass vessel of about one litre capacity, the fall of potential due to the natural ionization of the air inside the electroscope was about six volts per hour. The capacity of the gold-leaf system was about one electrostatic unit. The current is equal to the capacity multiplied by the fall of potential per second, *i. e.*,

$$\text{current} = \frac{1 \times 6}{3600 \times 300} = 5.6 \times 10^{-6} \text{ electrostatic units}$$

$$= 1.9 \times 10^{-15} \text{ amperes.}$$

With special precautions, a rate of discharge of $\frac{1}{10}$ of this amount can be accurately measured.

The number of ions produced in the electroscope can readily be deduced. J. J. Thomson has shown that an ion carries a charge of 3.4×10^{-10} electrostatic units or 1.13×10^{-19} coulombs. The number of ions produced per second in the air is thus,

$$\frac{1.9 \times 10^{-15}}{1.13 \times 10^{-19}} = 17000.$$

Supposing the ionization to be uniform throughout the volume of air, this corresponds to a production of 17 ions per c.c. per second in the volume of air inside the vessel of one litre capacity.

It will be shown later that the average α particle is able to produce about one hundred thousand ions in its path before it ceases to ionize the gas. We thus see that the electrical method is capable of detecting the ionization produced by a quantity of active matter which on an average expels one α particle per second; or, in other words, the electric method serves to detect a transformation of matter which takes place at the rate of one atom per second.

For the detection of matter which possesses the radioactive property, the electric method thus far transcends in delicacy the use of the spectroscope. In consequence of this, we are able to detect the presence of an active substance like radium, when it exists in almost infinitesimal amount mixed with inactive matter, and also to determine with fair accuracy the amount present. This test is so extraordinarily delicate that the presence of minute traces of radium has been observed in almost every substance which has been examined.

CHAPTER II

RADIOACTIVE CHANGES IN THORIUM

In the preceding chapter, a brief survey has been made of the more important properties possessed by the radioactive bodies, and a short description has been given of the various methods of measurement of radioactive quantities. In this and succeeding chapters, we shall analyze in more detail the processes occurring in the radiaoctive substances and the theories advanced for their explanation. Although in most of the subsequent chapters we shall discuss the properties of radium as our typical radioelement, yet the changes which occur in radium are so numerous and so complicated that it is advisable to consider a simpler illustration before entering upon the more complex problem.

For this purpose, we shall first consider the succession of changes taking place in thorium; for in that substance the analysis required is of a much simpler character than for radium itself. In this way we shall be able to concentrate our attention upon the main phenomena under consideration without being too much disturbed by details. When once the general principles involved have been made clear, their application to the more complex problem of radium will not present much additional difficulty.

This mode of beginning is also historically interesting, for it was as a result of the examination of the processes occurring in thorium that the disintegration theory of radioactivity, which will form the basis of the explanation of radioactive phenomena, was first outlined.

Thorium compounds have about the same radioactivity, weight for weight, as the corresponding compounds of uranium, and, like that substance, emit α, β, and γ rays. We have seen, however, that thorium differs from uranium in emitting, besides the three types of rays, an "emanation," or radioactive gas. This property possessed by thorium of emitting a volatile radioactive substance, can be readily illustrated by the simple experimental arrangement shown in Fig. 11.

A glass tube, A, is filled with about 50 grams of thorium oxide, or, still better, thorium hydroxide, which emits the emanation more freely than the oxide. This is connected by a narrow tube, L, of about a metre length, with a suitable, well insulated electroscope. On charging the electroscope, the leaves converge very slowly as the thorium compound has no appreciable effect in ionizing the gas inside the electroscope. If, however, a slow steady stream of air from a gas bag or gasometer

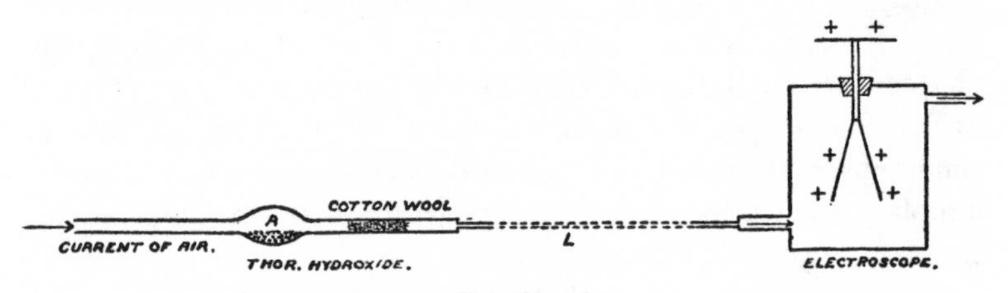

FIG. 11.

Discharging power of the thorium emanation.

is passed over the thorium, no effect is observed in the electroscope for a definite interval, which is a measure of the time taken for the air current to travel from A to the electroscope. The leaves are then seen to collapse rapidly, the rate of movement increasing for several minutes. This discharge of the electroscope is due to the ionizing effect of the thorium emanation, which has been conveyed with the current of air into the electroscope. On stopping the air current, the rate of collapse of the gold leaves is seen to diminish steadily, falling to half of its value in about one minute, or, to be more accurate, in 54 seconds. After about 5 minutes the effect of the emanation is almost inappreciable. The emanation, left to itself, loses its activity according to an exponential law. In the first 54 seconds the activity is reduced to half value; in twice that time, *i. e.* in 108 seconds, the activity is reduced to one quarter value, and in 162 seconds to one eighth value, and so on. This rate of decay of the activity of the thorium emanation is its characteristic feature, and serves as a definite physical method of distinguishing the thorium emanation from that of radium or of actinium which decay at very different rates.

Although the amount of emanation liberated from a kilogram of a thorium compound is far too minute to be detected either by its volume or weight, yet the electrical test of its presence is so extraordinarily delicate that its discharging effect can readily be observed with only a few milligrams of material.

The amount of emanation liberated into the air from a given weight of thorium varies enormously with the different compounds. For example, it is emitted freely by thorium hydroxide, but very slightly by thorium nitrate. Rutherford and Soddy [1] made a detailed examination of the "emanating power" of thorium compounds, *i. e.* the amount of emanation given off into the air per second by a given weight of material, and found that it was much affected by physical and chemical conditions. It is increased by the presence of moisture and by a rise of temperature up to a red heat. Lowering of the temperature to –80° C diminishes the emanating power considerably in many cases. The emanation is, however, given off freely, and to an equal extent by all compounds of thorium in solution. This is most readily shown by bubbling a current of air through the solution, when part of the emanation escapes mixed with the air. Rutherford and Soddy showed that the great variations in emanating power under different conditions were not due to differences in the rate of production of the emanation in the various cases, but merely *to variations in the rate of escape of the emanation into the air.* Since the thorium emanation loses a large proportion of its activity in a few minutes, any retardation in its rate of escape through the pores of a solid compound will very materially alter the emanating power. They concluded that, for equal weights of the element thorium, all thorium compounds produce an equal quantity of the emana-

[1] Rutherford and Soddy: Phil. Mag., Sept., 1902.

tion per second, but that the rate of its escape into the gas depends greatly on physical and chemical conditions.

We shall now briefly consider the chemical nature of the emanation itself, disregarding for the moment the substance from which it arises. Since the emanation is released in insignificant amount, no direct chemical examination can be made, but the conductivity produced in a gas by the emanation offers a very simple method of testing whether its amount is reduced after it has been acted on by various agents. For example, if the conductivity produced in a testing vessel is unaltered after passing the emanation slowly through a platinum tube at a white heat, and this is what has actually been observed, we can safely conclude that the emanation is unaffected by exposure to that temperature. In this way Rutherford and Soddy found that the thorium emanation was not acted on appreciably by any physical or chemical agent. The emanation was exposed to such severe treatment that no gas except one of the inert group of the argon-helium family could have possibly survived the various processes without change in amount. It was therefore concluded that the emanation is a chemically inert gas, which, in respect to the absence of any definite combining properties, is chemically allied to the argon-helium family.

The material nature of the emanation was strongly confirmed by the fact that it could be condensed from any gas with which it is mixed, by the action of extreme cold. The thorium emanation commences to condense from the air at a temperature of about –120° C. The emanation is in consequence completely stopped by passing the gas with which it is mixed, slowly through a U tube immersed in liquid air. From the rate of diffusion of the emanation through air and other gases, it has been deduced that the emanation is a gas of heavy molecular weight.

We have already seen that the emanation loses its activity rapidly according to an exponential law. If I_0 is the initial activity, the activity I_t at any time later is given by the equation

$$\frac{I_t}{I_0} = e^{-\lambda t},$$

where λ is a constant and e is the base of Naperian logarithms. Since the activity falls to half value in about 54 seconds the value of

$$\lambda = \frac{\log_e 2}{54} = .0128\,(\text{sec})^{-1}.$$

The decay curve of the thorium emanation is shown in Fig. 12. If the logarithm of the activity at any time is plotted with

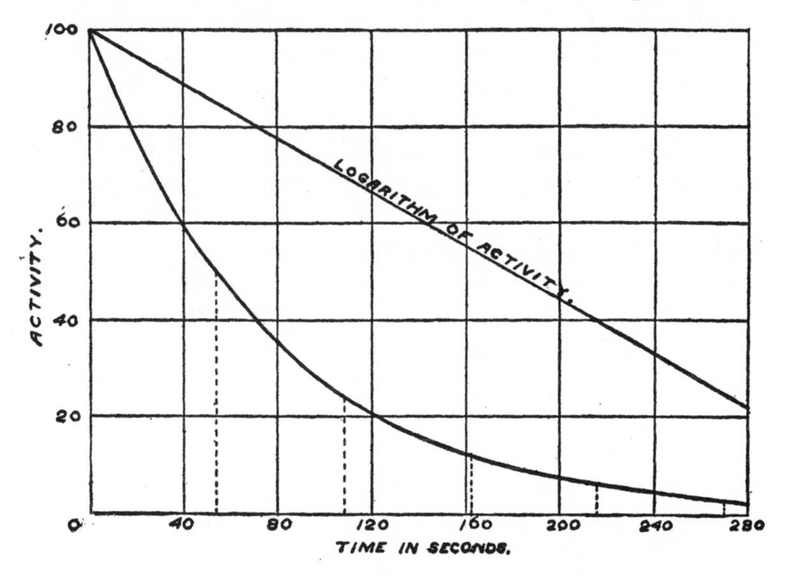

Fig. 12.

Decay of activity of the thorium emanation.

respect to time, the points all lie in a straight line. This is shown in the figure, the initial value of the logarithm of the activity being taken as 100 for convenience.

This value of λ is a characteristic constant of the emanation and will be termed the "radioactive constant." As far as observation has at present shown, its value is independent of physical or chemical conditions. For example, the value of λ is the same for the emanation when condensed by liquid air at a temperature of –186° C, as under normal conditions.

All simple radioactive products lose their activity according to an exponential law, and it is convenient to employ a single term to denote the time taken by any simple product to lose half of its activity. The term "*period*" of a product will be used in this sense to avoid circumlocution.

We must now consider the interpretation to be placed upon the observed law of decay of activity of the emanation. Is the activity of the emanation merely a transient superficial property, or is it directly connected with some essential change in the emanation itself? Consider for a moment how the activity is measured. The emanation gives out only *a* rays, which, as we have seen, are heavy positively charged particles projected with a speed of about twenty thousand miles per second. The ionization observed in the gas is due to the collision of the projected *a* particles with the gas molecules which lie in their path. The number of ions produced by each *a* particle under ordinary conditions is very great, probably amounting in some cases to about one hundred thousand. The activity measured by the electric method is thus a comparative measure of the number of *a* particles expelled from the emanation per second.

The *a* particles are apparently derived from the atoms of the emanation, and indeed it is difficult to avoid the conclusion that they were not projected initially from rest, but were in a state of rapid motion before their escape from the atom. It can be calculated that the *a* particle would have to move freely between two points differing in potential by about five million volts in order to acquire its enormous velocity of projection.

It is difficult to imagine any mechanism either within or outside the atom capable of suddenly setting in such rapid motion so heavy a mass as that of the *a* particle. We are almost forced to the conclusion that the *a* particle was originally in rapid motion within the atom and for some reason suddenly escaped from the atomic system with the velocity it originally possessed in its orbit. We may suppose that the expulsion of an *a* particle is a result of a violent explosion within the atom. The residual atom is lighter than before, and it is to be expected that its physical and chemical properties will be different from those of the parent atom. This, as we shall see later, is observed in all similar cases. As a result of the expulsion of α particles, the emanation of thorium is changed into another distinct substance which behaves like a solid and is deposited on the surface of bodies. This product of the decomposition, or rather disintegration, of the emanation will be discussed later.

If each atom of the emanation on breaking up expels one α particle, the observed law of decay of activity is expressed by the equation

$$\frac{n_t}{n_0} = e^{-\lambda t},$$

where n_0 is the initial number that breaks up per second, n_t is the number at any time t and λ is the radioactive constant. The same equation also holds if each atom in disintegrating expels two or more α particles.

Assuming for the sake of simplicity the probable law that the disintegration of each atom is accompanied by the appearance of one α particle, the number N_0 of atoms of emanation initially present must be equal to the total number of α particles expelled during the whole life of the emanation. This number is given by

$$N_0 = \int_0^\infty n_t \, dt = n_0 \int_0^\infty e^{-\lambda t} \, dt = \frac{n_0}{\lambda}.$$

The number N which remain unchanged after a period t is easily seen to be given by

$$N = \int_t^\infty n_t \, e^{-\lambda t} \, dt = \frac{n_0}{\lambda} e^{-\lambda t}.$$

Then

$$\frac{N}{N_0} = e^{-\lambda t}.$$

We have thus arrived at the important conclusion that the number of atoms of the emanation which remain unchanged at any time decreases in exactly the same way as the activity, or, in other words, the activity of the emanation is directly proportional to the number of atoms of emanation present. This is the result naturally to be expected from *a priori* considerations.

The exponential law of change of radioactive matter is the same as for a so-called monomolecular change in chemistry, observed in special cases when one of the two combining substances is present in a very large proportion compared with the other. The fact that the constant of decay is independent of the concentration of the emanation points to the conclusion that only one changing system is involved. The fact that the constant of decay does not depend on the physical and chemical conditions suggests that the changing system is the atom itself and not the molecule.

The radioactive constant λ has a definite physical meaning. We have seen above that

$$N = N_0 e^{-\lambda t}.$$

Differentiating with regard to the time,

$$\frac{dN}{dt} = -\lambda N,$$

or, the average number of atoms which break up per second is equal to the total number present multiplied by λ. The value of λ thus represents the *fraction* of the total number of atoms which disintegrate per second. For the thorium emanation in which half of the emanation breaks up in 54 seconds, the ratio $\lambda = .0128\ (\text{sec})^{-1}$. For example, suppose a vessel contained initially 10,000 atoms of the thorium emanation. In the first second, on the average, 128 atoms break up per second. After 54 seconds, the number of atoms of emanation present is 5000, and the number breaking up per second is 64. After 108 seconds the number of atoms of emanation left is 2500, and 32 break up per second, and so on. The value of λ has thus a definite and important physical meaning for any radioactive product.

We have spent some time in considering the physical interpretation to be placed upon the observed law of decay of activity for the emanation, since every radioactive product, so far examined, follows the same law of change, but with a different though definite value of λ, which is characteristic for each special type of radioactive matter. The same general explanation of the decay of activity may thus be directly applied to any radioactive product.

Excited Radioactivity of Thorium

The writer[1] showed that thorium compounds, besides emitting three types of rays and an emanation, possessed the follow-

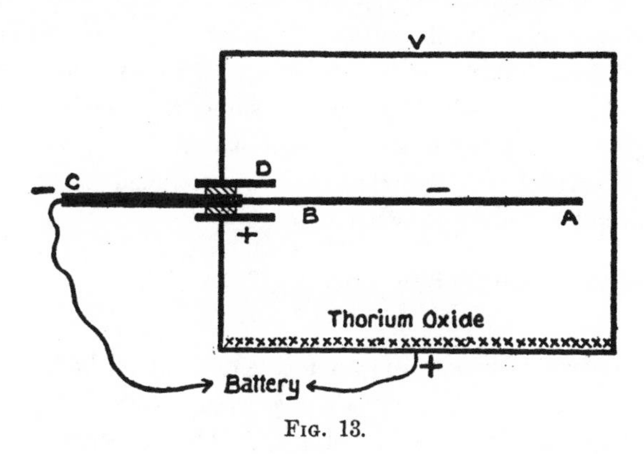

Fig. 13.

Concentration of the excited activity on the negative electrode.

ing remarkable property. Any body which has been exposed in the presence of the thorium emanation becomes itself radioactive. This "excited" or "induced" radioactivity, as it has been termed, is not permanent, but decays when the body is removed from the presence of the emanation. The activity can be largely concentrated on the negative electrode in a strong electric field. This can readily be shown by exposing a fine wire, AB, (Fig. 13) in the presence of an emanating thorium compound contained in a closed box, V.

When the wire is the only negatively charged body exposed to the emanation, it becomes intensely active. If it is charged positively, very little activity is observed. A fine wire may in

[1] Rutherford: Phil. Mag., Jan. and Feb., 1900.

this way be made many hundred times more active per unit area than thorium itself. The activity is due to a material deposit on the wire, for it can be partly removed by rubbing the wire with a piece of cloth, and can be dissolved off a platinum wire by strong acids. If the acid is evaporated in a dish to dryness, the active residue is left behind. The active matter can also be driven off by exposing the platinum wire to a temperature above a red heat. This property of the "*active deposit*," as it will be called, will be discussed in more detail later. The amount of activity under given conditions which can be concentrated on a body is independent of its chemical nature. Every substance made active in this way behaves as if it were coated with an invisible film of the same radioactive material. Although the active deposit is too small in quantity to be directly observed, the electrical effects produced by it are often large and very readily measured.

Connection between the Active Deposit and the Emanation

The property of producing an active deposit on bodies belongs not to thorium directly, but to the emanation emitted by it. The activity of the deposit, produced by exposure of a body near a thorium compound, depends on the emanating power of the compound. It is much greater, for example, for the hydroxide than for the nitrate, since the former emits emanation much more freely. If the thorium compound is completely covered by a very thin plate of mica which prevents the escape of the emanation, no excited activity is produced on a body placed outside it. This shows that the excited activity is not a consequence of some action of the rays directly emitted from thorium, since these are only slightly stopped by the mica, but is due to the presence of the emanation. The close connection between the emanation and excited activity is clearly brought out by the following experiment. A slow constant stream of air is passed over a large weight of thorium compound, and the stream of air mixed with emanation is then passed through a long tube, in which cylindrical electrodes, A, B, C, of equal length, are placed. The arrangement is clearly seen in Fig. 14. The conductivity of the gas, which is a measure of the amount of emanation present, falls off from electrode to electrode, since the emanation loses its activity with time. The ionization current observed, for example, between the electrode A and the outer cylinder is at first a measure of the amount of emanation present in the space between the cylinder and the electrode. After several hours, the stream of air was stopped, the central rods, A, B, C, were removed, and the excited activity determined by the electric method in an apparatus similar to that shown in Fig. 9. The excited activity was observed to fall off from electrode to electrode in exactly the same proportion as the

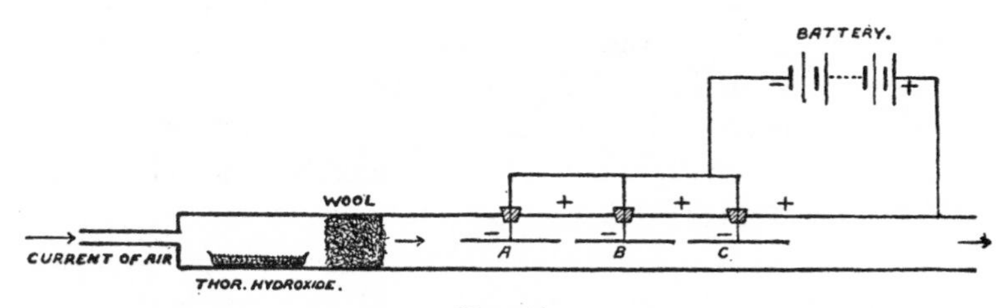

FIG. 14.

An experiment to show the connection between the emanation and the excited activity it produces.

activity due to the emanation alone. This shows that the excited activity produced is directly proportional to the amount of emanation present; for as the amount of emanation decreases, the excited activity falls off in the same ratio. This experiment also shows conclusively that the emanation causes the excited activity, since the latter is produced at points far removed from the action of the direct radiation from the thorium. The proportion that exists between the amount of the active deposit and of the emanation shows clearly that the latter is the parent of the former. It may be supposed that the residue of the atom of the emanation after an expulsion of the *a* particle becomes the atom of the active deposit. The new atom in some way gains a positive charge, and is conveyed to the negative electrode to which it adheres. In the absence of an electric field, the carriers of the active substance are conveyed by diffusion to the sides of the vessel containing the emanation. The number of particles of the active deposit produced per second in any space should be proportional to the number of particles of emanation which break up per second. The latter number, as we have seen, is proportional to the activity measured by the electrometer. The view that the particles of the active deposit result from the disintegration of the emanation thus leads at once to the conclusion that the amount of excited activity produced in any space is directly proportional to the activity of the emanation present, *i. e.* to the amount of emanation present. We may thus conclude with confidence that the active deposit is derived from the disintegration of the emanation, or, in other words, that the emanation is the parent of the active deposit.

There is a striking difference between the physical and chemical properties of the active deposit and of its parent the emanation. The latter, as we have pointed out, is a chemically inert gas, insoluble in acids, which condenses at about –120° C. The former behaves as a solid substance, is soluble in strong acids, and volatilizes at a temperature above a red heat. As a result of the disintegration of the emanation, a new substance is produced which differs completely both in physical and chemical properties from its parent.

Complexity of the Active Deposit

If a body is exposed for several days in the presence of the emanation, the excited activity after removal decays very nearly according to an exponential law, falling to half value in 11 hours. The active deposit emits α, β, and γ rays, and the rate of decay is the same whichever type of rays is used as a means of measurement. This at once suggests that a type of radioactive matter may be present which is half transformed in about 11 hours, emitting during the process the three types of rays. The active deposit is, however, more complex than this would indicate. If a body is exposed for only a few minutes to a large supply of emanation, the activity after removal varies in a very different way from that observed after a long exposure to the emanation. The activity is very small at first, but steadily increases for about 3.66 hours, when it reaches a maximum value. After six hours, the activity diminishes according to the eleven hour period, observed in the case of long exposures. The curve of decay is shown in Fig. 15.

This variation of the activity appears at first sight very remarkable and difficult to explain. The curve of variation of

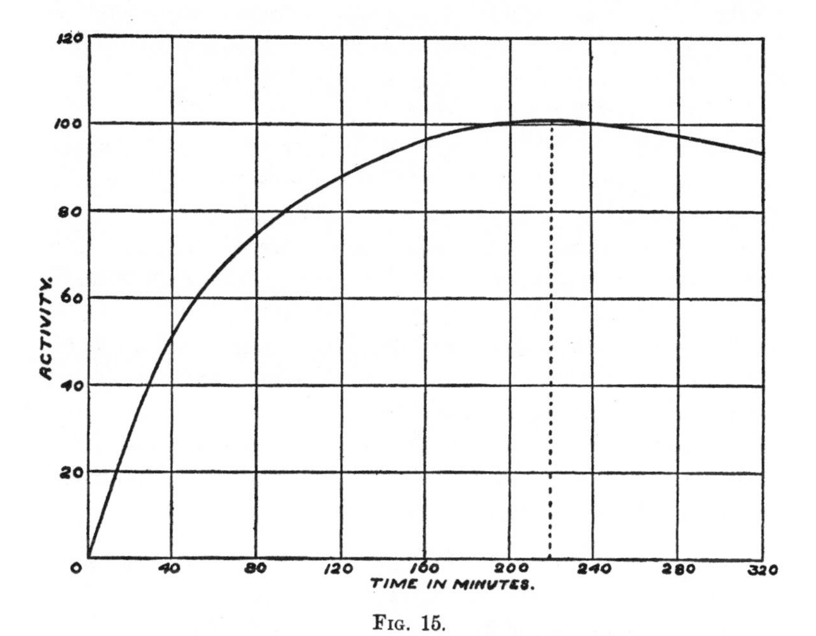

FIG. 15.

Variation with time of the excited activity produced after a short exposure to the thorium emanation.

activity is quite independent of the chemical nature of the body on which the active deposit is obtained. It is shown equally for deposits on thick plates of metal, and on the thinnest sheets of metal foil. The curve, however, can be completely explained if the active deposit contains two distinct substances, one of which is produced from the other. Let us suppose, for the moment, that the emanation changes into a substance called thorium A, and that after deposit on the surface of the body, thorium

A is gradually transformed into another distinct substance, thorium B. The substance thorium A is supposed to be transformed without the emission of either α, β, or γ rays, but thorium B emits all three types of rays. If the time of exposure to the emanation is very short compared with the period of transformation of thorium A or B, the matter deposited from the emanation will at first consist almost entirely of the inactive substance thorium A. The initial activity after removal will in consequence be very small. Since A gradually changes into B, and since the transformation of B is accompanied by the emission of rays, the activity will at first increase with the time, for more and more of B is produced. After a definite interval, the rate of transformation of B will exactly compensate for the rate of supply of B due to the change in A. At that moment the activity will be a maximum, and will afterwards decrease, since the amount of B will steadily diminish. This hypothesis is seen to account in a general way for the shape of the curve, but we shall now show that it also offers a complete quantitative explanation. Suppose that the constants of change of A and B are λ_1 and λ_2 respectively, and that n_0 atoms of A are deposited on the body during its short exposure to the emanation. After removal of the body, the number P of atoms of A diminishes according to the equation

$$P = n_0 e^{-\lambda_1 t}.$$

The rate of change of P is given by

$$\frac{dP}{dt} = -\lambda_1 n_0 e^{-\lambda_1 t}.$$

If Q is the number of atoms of B present at any time t later, the rate of increase of Q is equal to the rate of supply of fresh atoms in consequence of the transformation of A, less the rate of change of the atoms of B itself, *i. e.*,

$$\frac{dQ}{dt} = \lambda_1 n_0 e^{-\lambda_1 t} - \lambda_2 Q. \tag{1}$$

The solution of this equation is seen to be of the form

$$Q = a\,e^{-\lambda_1 t} + b\,e^{-\lambda_2 t};$$

and since $Q = 0$ when $t = 0$,

$$a + b = 0.$$

By substitution we find that

$$a = -\,b = \frac{\lambda_1\, n_0}{\lambda_2 - \lambda_1}.$$

Therefore, $$Q = \frac{\lambda_1\, n_0}{\lambda_2 - \lambda_1}\,(e^{-\lambda_1 t} - e^{-\lambda_2 t}).$$

The value of Q increases at first with the time, passes through a maximum, and then diminishes. A maximum is obtained when

$$\frac{dQ}{dt} = 0,$$

i. e., at a time T given by the equation

$$\frac{\lambda_2}{\lambda_1} = e^{-(\lambda_1 - \lambda_2)T}.$$

Since B alone gives out rays, the activity I_t at any time t is always proportional to the amount of B present: *i. e.*, to the value of Q. We therefore see that

$$\frac{I_t}{I_T} = \frac{e^{-\lambda_2 t} - e^{-\lambda_1 t}}{e^{-\lambda_2 T} - e^{-\lambda_1 T}}, \qquad (2)$$

where I_T is the maximum activity. Since the activity, whether for a long or short exposure, finally decays according to an exponential law with the period of 11 hours, it follows that either thorium A or thorium B is transformed according to this period. Let us suppose for the moment that half of A is transformed in 11 hours. The corresponding value of λ_1 is $1.75 \times 10^{-5}(\text{sec})^{-1}$.

It now remains for us to determine the period of B from consideration of the experimental curve. Since it is observed that the activity reaches a maximum value after a time $T = 220$ minutes, on substituting the values of λ_1 and T in equation (2), the value of λ_2 is found to be

$$\lambda_2 = 2.08 \times 10^{-4} \, (\text{sec})^{-1}.$$

This corresponds to a change in which half of the matter of thorium B is transformed in 55 minutes. On substituting these values of λ_1, λ_2, and T in equation (2), the value of $\frac{I_t}{I_T}$ can be at once determined. The very close agreement between the values deduced from the theory and the experimental numbers is shown in the following table.

Time in minutes.	Theoretical values of $\frac{I_t}{I_T}$.	Observed values of $\frac{I_t}{I_T}$.
15	.22	.23
30	.38	.37
60	.64	.63
120	.90	.91
220	1.00	1.00
305	.97	.96

The agreement is equally close for still longer periods. After about 6 hours the activity decreases very nearly exponentially, falling to about half value in 11 hours.

We thus see that a quantitative explanation of the activity curve can be obtained on the following assumptions:—

(1) That the matter thorium A deposited from the emanation is half transformed in 11 hours, but does not itself emit rays.

(2) That the matter thorium A changes into thorium B, which is half transformed in 55 minutes, and emits all three types of rays.

A very interesting point arises in the selection of the periods of the transformation of thorium A and B. We assumed that the period of 11 hours belonged to A rather than to B, but the activity curve itself gives us no information on this question. We see that equation (2) is symmetrical in regard to λ_1, λ_2, and in consequence would not be altered by an interchange of their values. In order to settle this question definitely, it is necessary to isolate thorium B from the mixture of A and B, and separately determine its period. If it is found possible to isolate an active product from the mixture of A and B which decays exponentially, falling off to half value in 55 minutes, it follows at once that the "ray" product B has this period, and that the 11 hour period belongs to A, the "rayless" product. This separation has actually been accomplished by several experimental methods, and the results completely confirm the theory already considered, and at the same time illustrate in a remarkable way the differences in physical and chemical properties of the two products, thorium A and B.

Pegram[1] examined the radioactivity produced in the electrodes by electrolysis of a thorium solution, and, under suitable conditions, obtained a product, the activity of which decayed exponentially, falling to half value in about 1 hour.

Von Lerch[2] made a number of experiments on the effect of electrolyzing a solution of the active deposit of thorium, obtained by solution in hydrochloric acid of the active deposit on a platinum wire. Deposits of varying rates of decay were obtained under different conditions, some decaying to half value in 11 hours, and others at a more rapid rate. By using nickel electrodes, he obtained an active substance which decayed exponentially, falling to half value in one hour. Considering the close agreement between the calculated and observed periods, viz., 55 and 60 minutes respectively, there can be no doubt that the ray product, thorium B, had been completely separated from the mixture of A and B by electrolysis. The rates of decay of the deposits obtained under different conditions are readily explained, for in most cases A and B are deposited electrolytically together, but in varying proportions.

This result was still further confirmed by Miss Slater,[3] using a different method. A platinum wire, made active by exposure

[1] Pegram: Phys. Rev., Dec., 1903.

[2] Von Lerch: Annal. d. Phys., Nov., 1903; Akad. Wiss. Wien, March, 1905.

[3] Miss Slater: Phil Mag., May, 1905.

to the emanation, was heated to a high temperature by means of an electric current. Miss Gates had previously[1] observed that although the activity of a platinum wire was lost by heating to a white heat, yet if the heated body was surrounded by a cold tube, the activity after heating was found to be distributed in undiminished amount over the interior of the tube. This experiment showed that the activity had not been destroyed by the action of a high temperature, but that the active matter had been volatilized by heat and redeposited on the surrounding cold bodies. Miss Slater examined the rates of decay, both of the activity left behind on the wire, and of that distributed on a lead cylinder surrounding the wire, after heating the latter for a short time at different temperatures. After exposure to a temperature of 700° C. for a few minutes, the activity of the wire was slightly reduced. The activity on the lead cylinder was small at first, but increased, reaching a maximum after about 4 hours, and then decaying exponentially with a period of 11 hours. This variation of activity is almost exactly the same as that observed (Fig. 15) for a wire exposed for a short interval to the thorium emanation; *i. e.*, under conditions in which the matter initially consisted almost entirely of thorium A. This result, then, shows that some thorium A was driven off by heat, and deposited on the surface of the lead cylinder. On heating to about 1000° C. nearly all the thorium A was removed, for it was observed that the activity left behind on the wire decayed exponentially, falling to half value in about 1 hour. At a temperature of 1200° C., nearly all the thorium B also was volatilized. These results thus show conclusively that the period of the ray product, thorium B, is about 1 hour, and that the period of 11 hours must be ascribed to the rayless product, thorium A. We therefore see that it has been found possible to isolate the components of a mixture of thorium A and B by two distinct methods, the one depending on the difference in electrolytic behavior of the two substances, and the other on their difference in volatility. This is a very interesting result, for it not only indicates the difference in physical and

[1] Miss Gates: Phys. Rev., p. 300, 1903.

chemical nature of the two components of the active deposit, but also shows how a separation of two substances existing together in almost infinitesimal amounts can be effected by specially devised methods.

It is at first sight a most surprising result that we are able, not only to detect the presence, but also to determine the physical and chemical properties of a product like thorium A, which does not manifest its presence by the emission of radiations. This, as we have seen, is rendered possible by the fact that the product of its transformation emits rays. But for this property the presence of thorium A or B would never have been detected by the means at our disposal.

It has been seen that after a long exposure to the thorium emanation, the excited activity at once commences to diminish. This result necessarily follows from general considerations. If the body is exposed in the presence of a constant supply of the emanation for about a week, the activity produced reaches a steady limiting value. When this is the case, the number of atoms of each product supplied per second is equal to the number of each which breaks up per second. Immediately after removal from the emanation, the amount of A begins to diminish according to an exponential law, and it can be shown both theoretically and experimentally that the activity, which is a measure of the amount of thorium B, does not at first diminish accurately according to an exponential law, with an 11 hour period, but somewhat more slowly. Several hours after removal, however, the decay is very nearly exponential.

It is a matter of interest to observe that the activity for a long exposure does not decay according to the period of the ray-emitting substance, but according to that of the "rayless" product. The decay in such a case will always follow the longer period, no matter whether the substance, which is transformed according to that period, gives out rays or not.

We may at this stage briefly summarize the conclusions arrived at:

(1) The thorium emanation is a gas which is half transformed in 54 seconds, and emits only α rays.

(2) The emanation changes into a solid called thorium A, which is half transformed in 11 hours, but which does not emit rays.

(3) The thorium A in turn changes into a product, thorium B, emitting α, β, and γ rays, which is half transformed in about 1 hour.

The successive changes occurring in the emanation are shown diagrammatically below:

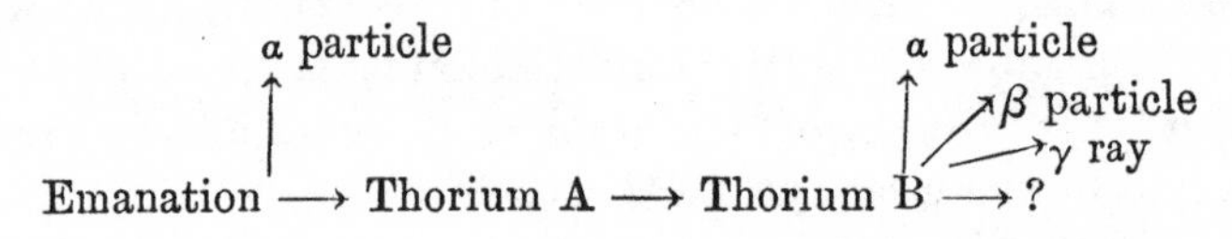

At present we have no definite information with regard to the product of transformation of thorium B. It is either inactive, or active to such a minute extent that its properties cannot be determined by the electric method.

Separation of Thorium X

It is now necessary to go back a stage and investigate the origin of the emanation. We shall first consider an important series of experiments by Rutherford and Soddy,[1] which have not only solved this question, but also have thrown a strong light on the processes occurring in thorium.

A small quantity of thorium nitrate was taken and dissolved in water. Sufficient ammonia was then added to precipitate the thorium present as hydroxide. The filtrate remaining was then evaporated to dryness, and the ammonium salts driven off by heating. The small residue finally obtained was, weight for weight, over one thousand times as active as the original thorium nitrate. The great activity of this residue, as compared with ordinary thorium, can be readily illustrated by means of the electroscope. The active residue obtained from 50 grams of the nitrate causes the gold leaves of the electroscope to collapse in a few seconds, while a weight of thorium nitrate equal to that of the residue causes hardly an appreciable movement.

[1] Rutherford and Soddy: Phil. Mag., Sept. and Nov., 1902.

The active substance present in this residue was called for convenience, thorium X (ThX). It probably exists in almost infinitesimal quantity mixed with the impurities left behind after the evaporation of the reagent, together possibly with a small trace of thorium which escaped precipitation. Since ThX is derived from the thorium salt, the latter must have been deprived of some of its activity. This was found to be the case, for the thorium hydroxide, so separated, had only about half of the activity to be normally expected.

The a ray activity of the ThX and the precipitated hydroxide were examined at intervals by means of an electrometer. The activity of ThX was not permanent, but increased for the first day and then decayed exponentially, falling to half value in about 4 days. After a month's interval, the activity sank to a minute fraction of its original value. The curve showing the variation of activity of ThX with time is seen in Fig. 16, Curve I.

Now let us turn our attention to the precipitated hydroxide. The activity of this decreased to some extent during the first day, passed through a minimum, and then steadily increased again with the time, reaching an almost steady value after a month's interval. These results are shown in Fig. 16, Curve II.

The two curves of decay of ThX, and of recovery of activity of the thorium, bear a very simple relation to one another. The initial rise in the ThX curve is seen to correspond to a fall in the recovery curve of the thorium, and when the activity of ThX has almost disappeared the activity of the thorium has practically reached a maximum value. The sum of the activities of the ThX and of the thorium from which it was separated is very nearly constant over the whole range of the experiment. The two curves of recovery and decay are complementary to each other. As fast as the ThX loses its activity, the thorium regains it. This relation between the curves is, at first view, most remarkable, and it would appear as if there were some mutual influence between the ThX and the thorium compound from which it was removed, so that the latter absorbed the activity lost by the former. This position is, however, quite untenable, for the rise and recovery curves are independent, and are unaltered if the thorium and ThX are kept in sealed vessels far removed from each other. If the thorium hydroxide after

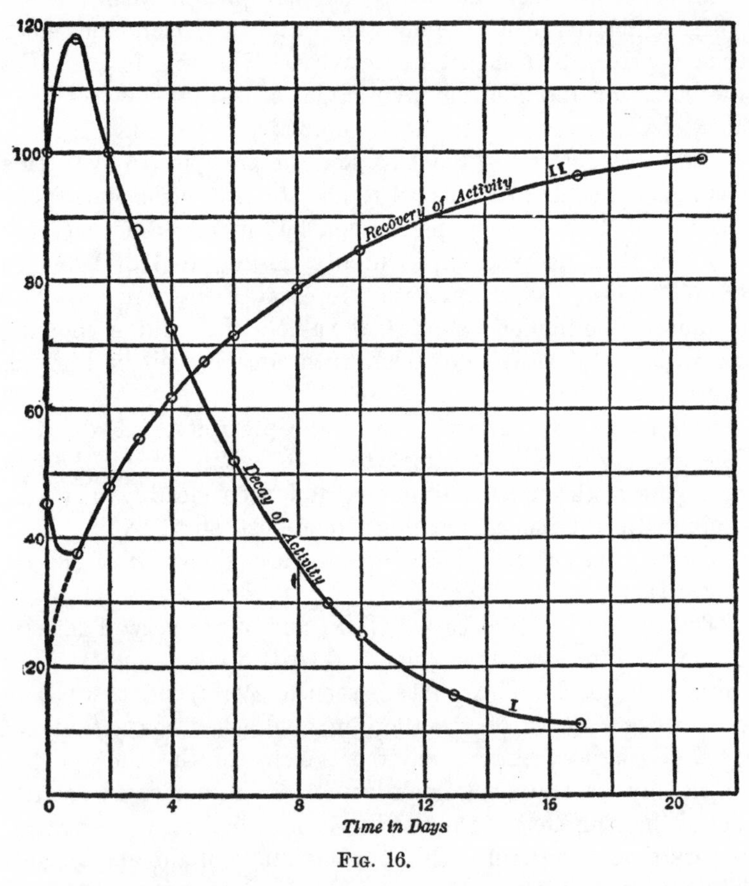

Fig. 16.

Decay of activity of thorium X and recovery of activity of thorium deprived of ThX.

recovering its activity is again dissolved and ammonia added, the amount of ThX separated is found to be the same as that obtained from the first experiment. This process can be repeated indefinitely, and equal quantities of ThX always be separated, provided that about a month elapses between each precipitation in order to allow the thorium to regain its lost activity. This shows that there is a fresh growth of ThX in the thorium after each precipitation.

We shall now consider the explanation of the connection between the decay and recovery curves. For the moment we shall disregard the initial irregularities shown by the two curves, which will be discussed later. If the recovery curve of Fig. 16

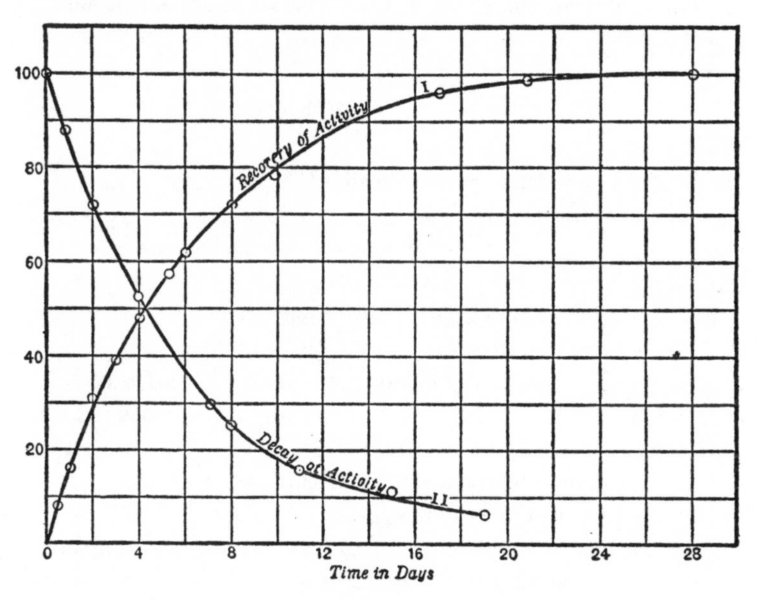

FIG. 17.

Decay curve of thorium X and recovery curve of thorium, from measurements one day after removal of the thorium X.

is produced backwards to cut the vertical axis, it does so at a minimum of 25 per cent. The curve of recovery of the lost activity reckoned from this 25 per cent minimum is shown in Fig. 17. In the same figure is shown the decay curve of ThX beginning after the second day and plotted to the same scale. The decay curve of the ThX is exponential, decreasing to half value in about 4 days. The decrease of activity from the initial value I_0 is given by the equation

$$\frac{I_t}{I_0} = e^{-\lambda t}.$$

The two curves are complementary, and the sum of the ordinates at any time is equal to 100 on the arbitrary scale. After 4 days, the activity of ThX has decayed to half value, and in the same interval the thorium has regained half its lost activity. The recovery curve is thus expressed by an equation of the form

$$\frac{I_t}{I_0} = 1 - e^{-\lambda t},$$

where I_t is the activity recovered after any time t, and I_0 the maximum activity which is regained when a steady state is reached. In this equation λ has exactly the same value as for the decay curve.

Following the same line of argument employed to interpret the decay curve of the emanation (page 43), we may suppose that the ThX is an unstable substance which is half transformed in 4 days, the amount of ThX breaking up per second being always proportional to the amount present. The radiation consisting of a rays accompanies the change, and is also proportional to the amount of ThX present.

Now we have seen that fresh ThX is produced in the thorium after the first supply has been removed. This production of ThX proceeds at a constant rate, but the amount of ThX present in the thorium cannot increase indefinitely, for at the same time the ThX is being changed continuously into another substance. A steady state will obviously be reached when the rate of production of new ThX exactly compensates for the rate of disappearance of ThX due to its own transformation. Now the number of atoms of ThX which break up per second is equal to λN where λ is the radioactive constant of ThX, and N is the number of atoms of ThX present at any time.

A steady state is reached when the number q of atoms of

fresh ThX supplied per second is equal to the number λN_0 which break up per second, where N_0 is the maximum number present when equilibrium is reached; *i. e.*,

$$q = \lambda N_0.$$

At any time the rate of increase $\frac{dN}{dt}$ of the number of atoms of ThX present is equal to the difference between the rate of supply and the rate of disappearance; *i. e.*,

$$\frac{dN}{dt} = q - \lambda N.$$

The solution of this equation is of the form

$$N = a e^{-\lambda t} + b,$$

where a and b are constants. Since $N = 0$ when $t = 0$, $a + b = 0$, and remembering that when $t = \infty$, N is equal to N_0,

$$a = -b = -N_0,$$

and consequently

$$\frac{N}{N_0} = 1 - e^{-\lambda t}.$$

This theoretical equation expressing the number of atoms of ThX present at any time is thus identical in form with the equation of variation of activity obtained experimentally. We therefore see that the decay and recovery curves of ThX are completely explained on the simple hypotheses: —

(1) That there is a constant production of fresh ThX from the thorium;

(2) That the ThX is continuously transformed, the amount changing per second being always proportional to the amount present.

The hypothesis (2) has been previously shown to be merely another method of expression of the observed exponential law of decay of the activity of ThX.

The first hypothesis can be proved experimentally. The

amount N of ThX present after the growth has been continued for a time t should be given by

$$\frac{N}{N_0} = 1 - e^{-\lambda t},$$

where N_0 is the equilibrium amount.

Since the ThX is half transformed in 4 days, $\lambda = .173\ (\text{day})^{-1}$. At the end of 1 day after complete removal of the ThX, the amount formed consequently should be 16 per cent of the maximum; after 4 days, 50 per cent; after 8 days, 75 per cent, and so on. Now it was found experimentally that three rapid precipitations of thorium by ammonia almost completely freed it from ThX for the time being. After standing for definite periods, the ThX present was removed and the amounts obtained were found to be in good agreement with the theory.

We thus see that the apparent constant radioactivity of thorium is really the result of two opposing processes of growth and decay; for radioactive matter is being continuously formed, and this matter in turn is continuously changing, and consequently losing its activity. There is thus a type of chemical equilibrium in which the rate of production of new matter balances the rate at which the new matter is transformed.

Source of the Thorium Emanation

A thorium compound completely freed from ThX gives off very little emanation, even in a state of solution. On the other hand, the ammonia solution which contains the ThX gives off a large amount. The removal of ThX is thus accompanied by the removal of the emanating power of thorium. It seems probable, therefore, that the emanation is derived from ThX, and further experiment has proved this to be the case. If a solution of ThX is taken, and a constant stream of air bubbled through it, the amount of emanation liberated is found to decrease exponentially, falling to half value in 4 days. This is exactly the result to be expected if ThX is the parent of the emanation, for the activity of ThX is a measure of the number of atoms of ThX breaking up per second, *i. e.*, a measure of the number of atoms of the new substance which is formed. The rate of production of emanation by the ThX should, on this view, be always proportional to the activity of ThX, and consequently should diminish at the same rate and according to the same law. This, as we have seen, has been experimentally observed. Although the thorium after removal of ThX is for the time almost entirely deprived of the power of emitting an emanation, this property is gradually regained, according to the same law as the recovery law of ThX shown in Fig. 17. This result follows naturally if ThX is the parent of the emanation. The emanating power should be proportional to the amount of ThX present, and should consequently vary *pari passu* with it.

We may thus conclude with confidence that the property of emitting an emanation is not a direct property of thorium itself, but belongs to its product ThX.

Initial Irregularities in the Decay and Recovery Curves

We are now in a position to explain the initial irregularities in the decay and recovery curves shown in Fig. 16. The activity of the separated ThX at first increases, while the activity of the precipitated thorium at first diminishes. Now the active deposit produced from the emanation is insoluble in ammonia, and consequently is left behind with the thorium. The ThX after separation produces the emanation, and this, in turn, is transformed into thorium A and B. The activity supplied by thorium B more than compensates at first for the decay of the activity of ThX alone. The activity consequently rises, but since the rates of transformation of A and B are rapid compared with that of ThX, after about one day equilibrium is practically reached, when very nearly the same number of atoms of ThX, and of each of its products, break up per second. When this is the case the activity of the emanation and of thorium B will vary exactly in the same way as that of the parent substance ThX. The activity of the active residue — which is a measure of the activity due to ThX, the emanation, and thorium B together — will in consequence decrease exponentially, falling to half value in four days.

Since the active deposit produced by the emanation in the mass of the thorium compound is not removed with the ThX, the activity due to it must at first diminish, for, in the absence of ThX and the emanation, there is no fresh supply of thorium A and B to compensate for their transformation. The activity of the thorium will thus diminish until the fresh supply of activity due to ThX and its succeeding products compensates for the decrease in the activity of the deposit. The activity will then be at a minimum, and will afterwards increase with the time, in consequence of the continued production of ThX.

The complementary character of the curves of decay and recovery, quite apart from the special considerations here advanced, is a necessary consequence of the laws governing radioactive changes. The rate of transformation, so far as observation has gone, is not affected by physical and chemical conditions. The transformation of ThX, when mixed with thorium, takes place at the same rate, and according to the same laws, as when it is isolated from the thorium by a chemical process. When the activity of a thorium compound has reached a constant value, the activity is then due to the various active products formed in it. If a product is separated by chemical or other means from the thorium compound, the activity due to this product plus that due to the thorium and the active products left behind, must be equal to the constant value of the activity of the original thorium in equilibrium. This follows at once, for otherwise there would be a creation or destruction of radioactivity by the mere removal of one of the products, and this would involve a gain or loss of radioactive energy. If, as in the case of ThX, the separated product first increases and then decreases in activity, there must be a corresponding decrease followed by an increase in the activity of the thorium from which it has been separated, in order that the sum of the activities of the two may be constant.

This principle of the conservation of the total amount of

radioactivity applies not only to thorium, but to radioactive substances generally. The total radioactivity of any substance in equilibrium cannot be altered by any physical or chemical agency, although the radioactivity may be manifested in a series of products, capable of separation from the parent substance. There is reason to believe, however, that the radioactivity of the primary active substances is not strictly permanent, but diminishes slowly, although in the case of feebly active elements like uranium and thorium, probably no appreciable change would be detected in a million years.

With an intensely active body like radium, it will be shown later that in all probability the sum total of the activity will ultimately decay exponentially, decreasing to half value in about thirteen hundred years. Provided, however, that the period of observation is small compared with the life of the primary substance, the principle of the constancy of the radioactivity is a sufficiently accurate expression of the experimental results. Many examples in support of this principle will be found in succeeding chapters of this book.

Methods of Separation of Thorium Products

In addition to ammonia, several reagents have been found capable of removing ThX from thorium solutions. Schlundt and R. B. Moore[1] found that pyridine and fumaric acid separate ThX from thorium nitrate solutions. These reagents differ from ammonia in removing the inactive product, thorium A, with the ThX, while the active product, thorium B, is left behind with the thorium.

Von Lerch[2] has shown that ThX can be separated by electrolysis from an alkaline solution of ThX, using amalgamated zinc, copper, mercury, or platinum as electrodes. The period of ThX has been accurately determined, and found to be 3.64 days. In addition, Von Lerch found that ThX was deposited on different metals by leaving them for several hours in an alkaline solution of ThX. Iron and zinc removed the greatest

[1] Schlundt and R. B. Moore: Journ. Phys. Chem., Nov., 1905.

[2] Von Lerch: Wien. Ber., March, 1905.

quantity. A nickel plate dipped into an acid solution of the active deposit becomes coated with thorium B, for the activity observed on the metal decays exponentially with a period of 1 hour. Other metals similarly treated also became active, but their rates of decay show that they are coated with a mixture of thorium A and of thorium B.

These results have shown in a striking way the differences in physical and chemical properties of the various thorium products. The methods of separation of the infinitesimal quantities of matter present are as definite as the ordinary chemical methods, applied to matter existing in considerable amount, while the radiating property serves as a simple and reliable method of qualitative and quantitative analysis.

Changes in Thorium

We have so far shown that thorium produces ThX, and that the latter is transformed into the emanation, which undergoes two further changes into thorium A and thorium B.

If thorium is subjected to a succession of precipitations with ammonia, extending over several days, the ThX is removed as fast as it is formed, and the active deposit has time to disappear. The activity of the thorium then sinks to a minimum of 25 per cent of its value when in equilibrium. The recovery curve of the thorium treated in this way does not show the initial decrease already referred to, but rises steadily, according to the recovery curve shown in Fig. 17. It is thus seen that thorium itself supplies only 25 per cent of the total a ray activity of thorium when in equilibrium, and that the rest is due to ThX, the emanation, and thorium B. Each of these a ray products supplies about 25 per cent of the total activity. Such a result is to be expected, for, when in equilibrium, an equal number of atoms of thorium ThX, the emanation, and thorium B must break up per second. This is based on the reasonable supposition, that each atom in breaking up gives rise to one atom of the succeeding product. The results, so far obtained, are completely explained on the disintegration theory put forward by Rutherford and Soddy. On this theory, a minute constant fraction of the atoms of thorium becomes unstable every second, and breaks up with the expulsion of an α particle. The residue of the atom after the loss of an α particle becomes an atom of a new substance, thorium X. This is far more unstable than the thorium itself, and breaks up with the expulsion of an α particle, half the matter being transformed in 4 days. ThX in turn changes into the emanation, which again breaks up into the active deposit, consisting of two successive products, thorium A and thorium B. The atom of thorium B breaks up with the accompaniment of an α and a β particle, and γ rays. Thorium A is transformed into thorium B without the appearance of rays. Such a change may consist either of a rearrangement of the parts constituting the atom without the projection of a part of its mass, or of the expulsion of an α particle at too low a velocity to ionize the gas. From the considerations advanced later in Chapter X, the latter supposition does not appear improbable.

A table of the products of thorium and some of their characteristic physical and chemical properties is given below.

TABLE OF TRANSFORMATION PRODUCTS OF THORIUM.

Radioactive product.	Time to be half transformed.	Nature of rays.	Some physical and chemical properties.
Thorium ↓	About 10^9 years	α	Insoluble in ammonia.
Thorium X ↓	4 days	α	Separated from thorium by its solubility in ammonia and in water and by electrolysis; separated also by fumaric acid and pyridine.
Emanation ↓	54 secs.	α	A chemically inert gas of high molecular weight; condenses from gases at a temperature of $-120°$ C.
Thorium A ↓	11 hours	No rays	Deposited on the surface of bodies; concentrated on the negative electrode in an electric field; soluble in strong acids; volatilized at high temperatures. A is more volatile than B. A can be separated from B by electrolysis and by its difference in volatility.
Thorium B ↓	1 hour	α, β, γ	
?	. . .	. . .	

The family of products of thorium is graphically represented in Fig. 18.

Radiothorium

There has been a considerable difference of opinion as to whether thorium is a true radioactive element or not, *i. e.*, as to whether the activity of thorium is due to thorium itself, or to some active substance normally always associated with it. Some experimenters state that by special methods they have obtained an almost inactive substance giving the chemical tests of thorium. Some recent work of Hahn[1] is of especial importance in this connection.

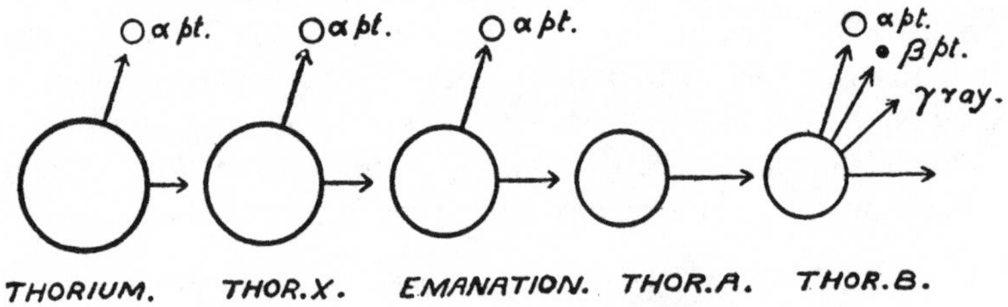

Fig. 18.

Family of thorium products.

Working with the Ceylon mineral, thorianite, which consists mainly of thorium and 12 per cent of uranium, Hahn was able by special chemical methods to separate a small amount of a substance comparable in activity with radium. This substance, which has been named "radiothorium," gave off the thorium emanation to such an intense degree that the presence of the emanation could be easily seen by the luminosity produced on a zinc sulphide screen. Thorium X could be separated from it in the same way as from thorium, while the excited activity produced by the emanation decayed with the period of 11 hours characteristic for thorium. The activity of radiothorium seems to be fairly permanent, and it seems probable that this active

[1] Hahn: Proc. Roy. Soc., March 16, 1905; Jahrbuch. d. Radioaktivität, II, Heft 3. 1905.

substance is in reality a lineal product of thorium intermediate between thorium and thorium X. The radiothorium produces thorium X, which in turn produces the emanation. It still remains to be shown that this active substance can be separated from ordinary thorium, but there can be little doubt that radiothorium is either the active constituent mixed with thorium, or, what is more probable, that it is a product of thorium. We shall see later that actinium itself is inactive, although it gives rise to a succession of active products remarkably similar in many respects to the family of products observed in thorium. The results of Hahn suggest that the transformation of thorium itself may be rayless, but that the succeeding product, radiothorium, gives out rays. Further results are required before such a conclusion can be considered as definitely established, but the results so far obtained by Hahn are of the greatest interest and importance.

CHAPTER III

THE RADIUM EMANATION

SHORTLY after the writer[1] had shown that thorium compounds continuously emit a radioactive emanation, Dorn[2] found that radium possesses a similar property. Very little emanation is emitted from radium compounds in a solid state, but it escapes freely when the radium is dissolved or heated. While the emanations of thorium and radium possess very analogous properties, they can readily be distinguished from each other by the difference in the rates of decay of their activities. While the activity of the thorium emanation decreases to half value in 54 seconds, and practically disappears in the course of 10 minutes, that of the radium emanation is far more persistent, for it takes nearly 4 days to be reduced to half value, and is still appreciable after a month's interval.

In physical and chemical properties the radium emanation is very similar to that of thorium, but, on account of its great activity and comparatively slow rate of change, it has been studied in more detail than the latter. It has been found possible to isolate it chemically and to measure its volume, as well as to observe its spectrum. The activity and concomitant heating effect, which are enormous in comparison with the amount of matter involved, have drawn strong attention to this substance, for the effects produced are of a magnitude that can neither be easily explained nor explained away. For these reasons, we shall consider in some detail the more important chemical and physical properties of the radium emanation, and the connection that exists between them. The study of this substance will throw additional light on the general theory of radioactivity which has already been developed in the last chapter.

[1] Rutherford: Phil. Mag., Jan., Feb., 1900.

[2] Dorn: Naturforsch. Ges. für Halle a. S., 1900.

The salts of radium, generally employed in experimental work, are the bromide and the chloride. Both of these compounds emit very little emanation into a dry atmosphere. The emanation produced is stored up or *occluded* in the mass of the substance, but is released by heating or dissolving the compound. The enormous activity of the emanation set free from radium is very well illustrated by the following simple experiment.

A minute crystal of the bromide or chloride is dropped into a small wash bottle. A few cubic centimeters of water are added to dissolve the compound, and the bottle is immediately closed. A slow current of air is then sent through the solution, and is carried along a narrow glass tube into the interior of an electroscope. If the electroscope is initially charged the leaves are observed to collapse almost immediately after the air reaches it. It is then found impossible to cause a divergence of the leaves for more than a moment. If the emanation is all blown out from the electroscope by a current of air, the leaves are still observed to collapse rapidly, although the emanation has been completely removed.

This residual activity is due to an active deposit left on the sides of the vessel. In this respect, the emanation of radium possesses a similar property to that of thorium. The activity, however, diminishes more rapidly than in the case of thorium, for most of the electrical effect due to it disappears in a few hours, while in the case of thorium the effect lasts for several days.

Measurements of the rate of decay of the activity of the emanation have been made by several observers. Rutherford and Soddy[1] stored a quantity of air mixed with emanation in a small gasometer over mercury, and a definite volume was withdrawn at intervals and discharged into a testing vessel such as is shown in Fig. 10. The activity observed in the vessel increased for several hours after the introduction of the emanation on account of the formation of the active deposit. By determining the saturation current immediately after the passage of the emanation into the testing cylinder, the quantity of emanation initially

[1] Rutherford and Soddy: Phil. Mag., April, 1903.

present was measured. In this way it was found that the amount of emanation present decreased according to an exponential law, falling to half value in 3.77 days. P. Curie [1] determined the constant of decay of the emanation in a somewhat different way. A large quantity of emanation was introduced into a glass tube, which was then sealed off, and the ionization due to the issuing rays was measured at intervals by an electrometer in a suitable testing vessel. Now it will be seen later that the emanation gives out only α rays, which are completely stopped by a thickness of glass less than $\frac{1}{10}$ mm.; consequently the rays from the emanation were absorbed in the walls of the glass tube. The electrical effect produced in the testing vessel was due entirely to the β and γ rays which are emitted from the active deposit produced on the inside of the tube by the emanation. Since after about 3 hours the active deposit is in radioactive equilibrium with the emanation, and then decays at the same rate as the parent substance, the intensity of the β and γ rays will diminish at the same rate and according to the same law as the emanation itself. In this way the activity was found to diminish according to an exponential law, falling to half value in 3.99 days. The agreement of these periods of decay obtained by different methods shows that the amount of the active deposit is always proportional to the amount of emanation present at any time during the life of the emanation. This is one of the proofs that the active deposit is a product of the decomposition of the emanation.

Further experiments to determine the constant of decay of the emanation have been made by Bumstead and Wheeler,[2] and Sackur.[3] The former found that the activity decreased to half value in 3.88 days, and the latter found the period to be 3.86 days. We may thus conclude that the emanation decays exponentially with a period of about 3.8 days.

The emanation from radium is almost entirely released by boiling a solution of the compound or by aspirating air through it. The active deposit is left behind with the radium, but this

[1] P. Curie: Comptes rendus, cxxxv, p. 857 (1902).

[2] Bumstead and Wheeler: Amer. Jour. Science, Feb., 1904.

[3] Sackur: Ber. d. d. chem. Ges., xxxviii, No. 7, p. 1754 (1905).

disappears after several hours. If the radium solution is then evaporated to dryness, the activity measured by the α rays is found to have reached a minimum of about 25 per cent of the normal value. If kept in a dry atmosphere, the emanation produced from the radium is occluded in its mass, and the activity of the radium consequently increases, reaching its normal steady value after about one month. The recovery curve of the activity of radium from the 25 per cent minimum is shown in Fig. 19. The decay curve of the emanation is added for comparison.

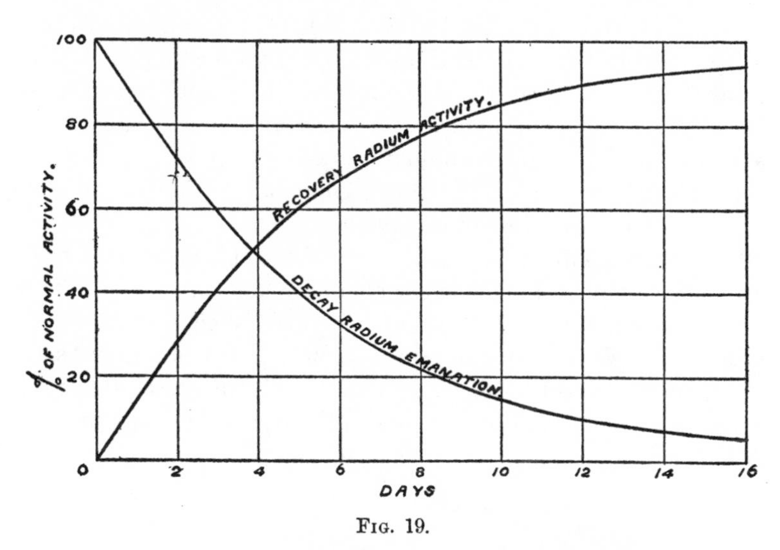

Fig. 19.

Decay curve of the radium emanation and recovery curve of the activity of radium measured by the α rays, from the 25 per cent minimum.

As in the case of thorium, the decay and recovery curves are complementary to each other. The activity of the emanation falls to half value in about 3.8 days, while half of the lost activity of the radium is recovered in the same interval.

The activity of the emanation released from the radium is thus given at any time by the equation

$$\frac{I_t}{I_0} = e^{-\lambda t},$$

while the equation of the recovery curve from the minimum is

$$\frac{I_t}{I_0} = 1 - e^{-\lambda t};$$

i. e., the amount of the emanation N stored up in the radium after standing for a time t is given by

$$\frac{N_t}{N_0} = 1 - e^{-\lambda t},$$

where N_0 is the maximum amount. These curves are explained in exactly the same way as the similar curves for thorium. The emanation is an unstable substance which is half transformed in 3.8 days. It is produced at a constant rate by the radium, and the activity of the radium reaches a steady value when the rate of production of fresh emanation balances the rate of disappearance of that already formed.

A radium compound initially freed from emanation will have grown a maximum supply again about one month later, and this process of removal and fresh growth may be continued indefinitely. If N_0 be the number of atoms of emanation present when in equilibrium, the rate q of supply of fresh atoms of emanation by the radium is equal to the number lost by its own decomposition, *i. e.*,

$$q = \lambda N_0, \qquad \lambda = \frac{q}{N_0}.$$

The value of λ thus has a definite physical meaning, for it represents the fraction of the equilibrium amount of emanation supplied per second, as well as the fraction of the emanation which breaks up per second. Taking the period of the emanation as 3.8 days, the value of λ with the second as the unit of time is 1/474000, or, in other words, the rate of supply of the emanation per second is 1/474000 of the equilibrium amount.

This result is well illustrated by a very simple experiment described by Rutherford and Soddy. A small quantity of radium chloride in radioactive equilibrium was dropped into hot water. The accumulated emanation released by solution was swept with a current of air into a suitable testing vessel, and the saturation current immediately measured. The current so determined is a comparative measure of N_0, the equilibrium amount of emanation stored up in the radium.

The radium solution was then aspirated with air for some time, to remove the last trace of accumulated emanation, and then allowed to stand undisturbed for 105 minutes. The emanation accumulated in this interval was then swept into a similar testing vessel and the saturation current again determined. This current is a measure of the amount N_t of the emanation formed in the interval. The ratio $\frac{N_t}{N_0}$ was found to be .0131, and disregarding the small decay of the emanation during such a short interval,

$$N_t = q \times 105 \times 60.$$

It follows that
$$\frac{q}{N_0} = 1/480000.$$

Allowing for the small decay during the interval,

$$\frac{q}{N_0} = 1/477000.$$

From the constant of decay of the emanation we have seen that

$$\lambda = \frac{q}{N_0} = 1/474000.$$

The agreement between theory and experiment is thus remarkably close, and is a direct proof that the production of emanation in a solid compound proceeds at the same rate as in the solution. In the former case it is occluded, and in the latter, part is retained in the solution and the rest in the air space above it.

It is surprising how tenaciously the emanation is held by dry radium compounds. Experiment showed that the emanating power in the solid state was less than one half per cent of the emanating power in solution. Since a radium compound stores up nearly 500,000 times as much emanation as is produced per second, the result shows that the amount of emanation escaping per second is less than one hundred millionth part of that occluded in the compound. The rate of escape of emanation is much increased in a moist atmosphere and by rise of temperature.

The recovery curve of a solid radium compound freed from emanation is altered if the conditions allow much of the recovered emanation to escape. Under such conditions, the maximum activity is reached more quickly, and is far smaller than the normal activity of a non-emanating compound.

This property of radium of retaining its emanation is difficult to explain satisfactorily unless it is assumed that there is some slight chemical combination between the emanation and the radium producing it. Godlewski[1] has suggested that the emanation is in a state of solid solution with the parent matter. This point of view is supported by certain observations made by him on the rapidity of diffusion of the product uranium X into a uranium compound. A discussion of his results will be given later in Chapter VII.

Condensation of the Emanation

For several years after the discovery of the emanations from thorium and radium, there existed considerable difference of opinion as to their real nature. Some physicists suggested that they were not material, but consisted of centres of force attached to the molecules of gas with which the emanation was mixed, and moving with them. Others held that the emanation was a gas present in such minute amount that it was difficult to detect by means of the spectroscope or by direct chemical methods. The objections urged against the material character of the emanation were to a large extent removed by the discovery, made by Rutherford and Soddy,[2] that the emanations of thorium and radium possessed a characteristic property of gases inasmuch as they could be condensed from the inactive gas with which they were mixed by the action of extreme cold.

[1] Godlewski: Phil. Mag., July, 1905.

[2] Rutherford and Soddy: Phil. Mag., May, 1903.

As a result of a careful series of experiments, it was found that the emanation from radium condensed at a temperature of –150° C. The condensation and volatilization points were very sharply defined, and did not differ by more than 1° C. The thorium emanation commenced to condense at about –120° C., but the condensation was not usually completed until a temperature of –150° C. was reached. The probable cause of this interesting difference in behavior of the two emanations will be discussed presently.

If a large amount of emanation is available, the condensation of the radium emanation can readily be followed by the eye. The experimental arrangement is clearly shown in Fig. 20.

The emanation mixed with air is stored in a small gasometer, and is then slowly passed through a U tube immersed in liquid air. This U tube is filled with fragments of willemite, or crystals of barium platinocyanide, which become luminous under the influence of the rays from the emanation. If the current of air mixed with emanation is passed very slowly through the tube, the fragments of willemite begin to glow brightly just below the level of the liquid air, and the luminosity can be concentrated over a short length of the tube. This shows that the emanation has been condensed at the temperature of liquid air, and is deposited on the walls of the tube and on the surface of the willemite. If the U tube is then partially exhausted and closed with stopcocks, the emanation still remains concentrated for some minutes on the tube and willemite, although the liquid air is removed. When, however, the temperature of the tube rises to –150° C., the emanation is rapidly volatilized, and distributed throughout the tube. This

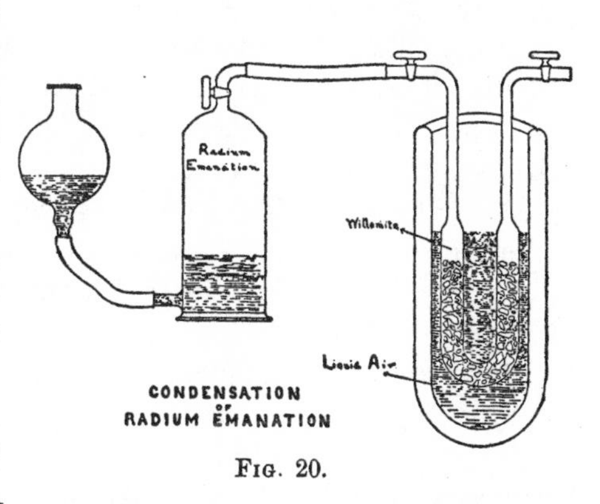

FIG. 20.

is observed by the sudden distribution of the luminosity throughout the whole mass of willemite in the U tube. The point of condensation remains brighter than the rest of the tube for some time. This is due to the fact that the emanation, even in the condensed state, has produced the active deposit. When the emanation is volatilized, the active deposit remains behind, and the rays from it cause a greater luminosity at that point. After an hour's interval this difference of luminosity has almost disappeared, and the willemite glows throughout with a uniform light. The luminosity can at any time be concentrated at any point by local cooling with liquid air.

If the U tube is filled with different layers of phosphorescent materials, like willemite, kunzite, zinc sulphide, and barium platinocyanide, the emanation after volatilization is equally distributed, and each layer of material glows with its own peculiar light. The greenish luminosity of the willemite and barium platinocyanide is not easily distinguishable, except for a difference of intensity. The kunzite glows with a deep red color, while the zinc sulphide emits a yellow light. There are several interesting points of distinction between the action of the rays of the emanation and of the active deposit on these substances. Unlike the other substances mentioned, the luminosity of zinc sulphide largely disappears at the temperature of liquid air, but revives at a higher temperature. The α rays produce a marked luminosity in willemite, the platinocyanides, and zinc sulphide, but have little or no effect in lighting up kunzite. The latter is sensitive only to the β and γ rays emitted from the active deposit. In consequence of this, the kunzite is very feebly luminous when the emanation is first introduced. The light, however, increases in intensity as the active deposit is produced by the emanation, and reaches a maximum about three hours after the introduction of the emanation. After barium platinocyanide has been exposed for some time to the action of a large amount of the emanation, the crystals change to a reddish tinge, and the luminosity is much reduced. This has been shown to be due to a permanent change in the crystals by the action of the rays. By re-solution and crystallization, the luminosity again returns.

Curie and Debierne early showed that glass becomes luminous under the action of the rays from the emanation. This effect is most marked in Thuringian glass, but as a rule the luminosity is feeble compared with that produced in willemite or zinc sulphide. The glass becomes colored under the action of the rays, and with strong emanation is rapidly blackened.

The sharpness of the temperature of volatilization of the radium emanation was very clearly illustrated by some experiments made by Rutherford and Soddy, using the electric method.

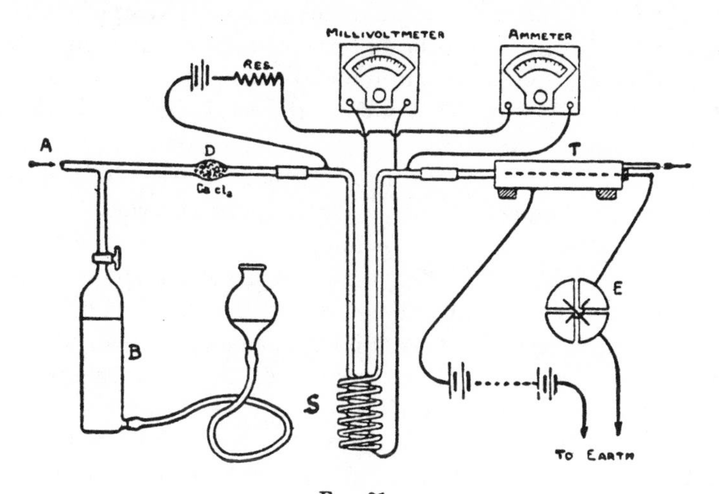

FIG. 21.

Determination of the temperature of condensation of the radium emanation by the electric method.

The emanation collected in the gasometer, B, was condensed in a long spiral copper tube, S, (see Fig. 21) immersed in liquid air, and a slow steady stream of air after passing through the tube entered a small testing vessel, T. After condensation the copper spiral was removed from the liquid air and allowed to heat up very slowly. The temperature was deduced from measurements of the resistance of the copper spiral. Just before the point of volatilization was reached, very little effect was observed in the testing vessel. Suddenly a rapidly increasing movement of the electrometer needle was noted, and by using a large quantity of emanation the rate of movement increased in a few moments from several divisions to several hundred divisions per second. The rise of temperature observed between the point at which there was practically no escape of the emanation and the point of rapid escape was not more than a fraction of a degree in many cases.

It has been already pointed out that the temperature of complete condensation of the thorium emanation is not at all sharp, but that the condensation in most cases continues over a range of about 30° C. This striking difference in the behavior of the two emanations is, in all probability, due to the small amount of the thorium emanation present in the experiments. The emanation of thorium breaks up at about six thousand times the rate of that of radium. For an equal expulsion of *a* particles by the two emanations, *i. e.* for approximately equal electrical effects, the latter must therefore be present in at least six thousand times the amount of the former. In addition, in most of the experiments with the radium emanation, the quantity of emanation was sufficient to produce several hundred times the electrical effect observed with the small quantity of the emanation obtained from thorium compounds. Thus, in some of the experiments, the quantity of radium emanation present was at least ten thousand times — and in many cases more than a million times — the amount of the thorium emanation. In fact, it can readily be calculated that in the actual experiments not more than 100 atoms of thorium emanation could have been present per cubic centimetre of gas carried through the copper spiral. Under such conditions, it is not so much a matter of surprise that the emanation of thorium does not show a sharp condensation point, as that the emanation can be condensed at all when so sparsely distributed throughout a volume of gas.

Diminution of the pressure of the air in the spiral, or the substitution of hydrogen for oxygen as the carrying medium, both tended to cause more rapid condensation. Such an effect is to be expected on the above view, since the rapidity of diffusion of the atoms of emanation through the gas is thereby hastened.

If the thorium emanation should ever be obtained in large

quantity, there can be little doubt that it will also exhibit comparatively sharp points of condensation and volatilization. The fact that the thorium emanation begins to condense at a higher temperature (–120° C.) than the radium emanation (–150° C.) shows that the emanations consist of different types of matter. The emanation of actinium, like the emanations from radium and thorium, may be condensed by passing it through a spiral immersed in liquid air, but the rapidity of the decay of its activity (half value in 3.9 seconds) makes an accurate determination of its condensation temperature by the electric method very difficult, since the emanation would lose the greater part of its activity before the stream of gas carrying the emanation could be reduced to the temperature of the spiral. The ease with which the radium emanation is condensed by liquid air has proved of great importance in many recent researches on the emanation. By the use of this property, it has been freed from the gases mixed with it, isolated in a pure state, and its spectrum determined.

Rate of Diffusion of the Emanation

If the emanation is introduced at one end of a tube kept at constant temperature, after the lapse of several hours it is found to be distributed in equal amount throughout the volume of the tube. This result shows that the emanation diffuses through the air like an ordinary gas. It has not yet been found possible to determine by a direct method the density of the emanation, as the quantity released from even one gram of pure radium bromide would be too small to be weighed accurately. By comparing the rate of diffusion of the emanation with that of a known gas, we can, however, obtain a rough estimate of its molecular weight. The rates of interdiffusion of various gases have long been known to decrease with the molecular weight of the diffusing gas. If therefore, for example, we find that the coefficient of interdiffusion of the emanation into air lies between the corresponding values obtained for two known gases, A and B, it is probable that the molecular weight of the emanation is intermediate in value between that of A and B.

Shortly after the discovery of the radium emanation, Rutherford and Miss Brooks[1] determined its coefficient of interdiffusion K into air, and found values lying between $K = .07$ and $K = .09$. The method adopted was to divide a long cylinder into equal parts by a movable slide. The emanation was first introduced into one half of the tube, and thoroughly mixed with the air. When temperature conditions were steady, the slide was opened, and the emanation gradually diffused into the other half. The amount of emanation present in each half of the tube, at any time after opening the slide, was determined by the electric method, and from these data the coefficient of interdiffusion can be calculated. The coefficient of interdiffusion of carbon dioxide (molecular weight 44) into air was long ago found to be .142. The emanation thus diffuses into air more slowly than does carbon dioxide into air. For alcohol vapor (molecular weight 77), the value of $K = .077$. Taking the lower value, $K = .07$, as the more probable value for the radium emanation, it follows that the emanation has a molecular weight greater than 77.

A number of determinations have since been made, by different methods, to form an estimate of the molecular weight of the emanation.

Bumstead and Wheeler[2] measured directly the comparative rates of diffusion of the emanation and of carbon dioxide through a porous pot. Assuming Graham's law, viz., that the coefficient of interdiffusion is inversely proportional to the square root of the molecular weight, they deduced that the molecular weight of the emanation was about 172.

Makower,[3] using a similar method, compared the rates of diffusion of the radium emanation through a porous pot with the rates for the gases oxygen, carbon dioxide, and sulphur dioxide, and finally concluded that the emanation had a molecular weight in the neighborhood of 100. Curie and Danne[4] determined the

[1] Rutherford and Miss Brooks: Trans. Roy. Soc., Canada, 1901; Chemical News, 1902.

[2] Bumstead and Wheeler: Amer. Journ. Sci., Feb., 1904.

[3] Makower: Phil. Mag., Jan., 1905.

[4] Curie and Danne: Comptes rendus, cxxxvi, p. 1314 (1904).

rate of diffusion of the emanation through capillary tubes, and obtained a value $K = .09$, a value somewhat higher than that obtained by Miss Brooks and the writer.

It is thus seen that all the experiments on diffusion bear out the conclusion that the emanation is a heavy gas with a molecular weight probably not less than 100. It is doubtful, however, whether much reliance can be placed on the actual value of the molecular weight deduced in this way, because the emanation exists in minute amount in the gas in which it diffuses, and its coefficient of interdiffusion is compared with that of gases existing in large quantity. The coefficients of interdiffusion may not in such a case be directly comparable. In addition, the rate of diffusion of the emanation, which has the properties of a monatomic gas, is compared with the rates of diffusion of gases which have complex molecules.

If the emanation is considered to be a direct product of radium, and to consist of the radium atom minus one or two α particles, the molecular weight should be not much less than the atomic weight of radium, viz., 225. It is doubtful whether the value of the molecular weight of the emanation can be determined with any certainty until the emanation has been obtained in sufficient amount to determine its density.

The coefficient of interdiffusion of the thorium emanation into air has been determined by the writer to be about .09. This would suggest that the thorium emanation has a somewhat smaller molecular weight than that of radium.

The emanation obeys the laws of gases, not only as regards diffusion, but also in other particulars. For example, the emanation divides itself between two connected reservoirs in proportion to their volumes. P. Curie and Danne showed that if one of the reservoirs was kept at a temperature of 10° C. and the other at 350° C., the emanation is distributed between them in the same proportion as a gas under the same conditions.

The emanation thus possesses the characteristic properties of gases, namely, condensation and diffusion. It also obeys at low temperatures Charles's law, and, as will be seen later, Boyle's law.

We may thus conclude with confidence that the emanation, while it exists, is a radioactive gas of heavy molecular weight.

Physical and Chemical Properties of the Emanation

A number of experiments have been made to determine whether the emanation possesses any definite chemical properties which would enable us to compare it with any other known gas, but so far no evidence has been obtained that the emanation is able to combine with other substances. In such experiments, the electric method offers a simple and accurate method of determining whether the quantity of the emanation is reduced under various conditions. In fact, it serves as a rapid and exact method of quantitative analysis of the minute amount of emanation under experiment.

Rutherford and Soddy[1] showed that the emanation was not diminished in quantity after condensation by liquid air, or by passage through a platinum tube kept at a white heat by an electric current. A number of experiments were also made in which the emanation was made to pass over a number of reagents, the emanation being always mixed with a gas unaffected by the particular reagent. They concluded from these experiments that no gas could have survived in unaltered amount the severe treatment to which it had been exposed, except an inert gas of the helium-argon family.

Ramsay and Soddy[2] found that the quantity of emanation was unchanged after sparking for several hours with oxygen over alkali. The oxygen was then removed by ignited phosphorus, and no visible residue was left. Another gas was then introduced, and the emanation after mixture with it was withdrawn. Its activity was practically unaltered. A similar result was observed when the emanation was introduced into a magnesium lime tube which was heated for three hours to a red heat.

[1] Rutherford and Soddy: Phil. Mag., Nov., 1902.

[2] Ramsay and Soddy: Proc. Roy. Soc., lxxii, p. 204 (1903).

We may thus conclude that the radium emanation in respect to the absence of definite combining properties is allied to the recently discovered inert gases of the atmosphere. On the disintegration theory, the emanation is supposed to be transformed with the accompaniment of the expulsion of α particles. It is of great importance to settle whether the rate of disintegration is affected by temperature. Any change in the rate of transformation would result in a change in the period of decay of the emanation. This point has been examined by P. Curie, who found that the decay of activity was unaffected by continued exposure of the emanation to temperatures varying between –180° C. and 450° C.

This result shows that the transformation of the emanation cannot be considered to be a type of ordinary chemical dissociation, for no reaction is known in chemistry which is independent of temperature over such a wide range. In addition, the transformation of the emanation is accompanied by the expulsion of a portion of its mass at enormous speed — a result never observed in chemical reactions. Such a result suggests that the change that occurs is not molecular but atomic. This view is strongly confirmed by the enormous release of energy during the disintegration of the emanation which will be considered later.

Volume of the Emanation

It has been seen that the amount of emanation to be obtained from a given quantity of radium reaches a maximum value when the rate of supply of fresh emanation balances the rate of transformation of that already produced. Since this maximum amount of emanation is always proportional to the quantity of radium present, the volume of emanation released from one gram of radium in radioactive equilibrium should have a definite constant value. It was early recognized that the volume of the emanation to be obtained from one gram of radium was very small, but not too minute to be measured. From the data available at the time, the writer[1] in 1903 calculated that the volume

[1] Rutherford: Nature, Aug. 20, 1903; Phil. Mag., Aug., 1905.

of the emanation derived from one gram of radium probably lay between .06 and .6 cubic millimetres at atmospheric pressure and temperature.

A more accurate deduction can be made from the more recent experimental data of the number of α particles expelled from one gram of radium per second. This number has been determined by the writer by measuring the positive charge communicated to a body on which the α rays impinged. Assuming that each α particle carries an ionic charge of 3.4×10^{-10} electrostatic units, it was deduced that one gram of radium at its minimum activity (*i. e.*, when the emanation and its disintegration products were removed) emitted 6.2×10^{10} α particles per second. If we suppose, as is probably the case, that each radium atom in breaking up gives rise to one atom of the emanation, the number of atoms of emanation produced per second is equal to the number of α particles expelled per second.

But N_0, the maximum number of atoms of emanation stored up in radium in radioactive equilibrium, is given by $N_0 = \frac{q}{\lambda}$, where q is the rate of production and λ is the decay constant.

Consequently, the value of N_0 for one gram of radium is $6.2 \times 10^{10} \times 474{,}000$, or 2.94×10^{16}.

Now from experimental data it is known that one cubic centimetre of any gas at atmospheric pressure and temperature contains 3.6×10^{19} molecules. Assuming that the molecule of the emanation consists of one atom, the volume of emanation from one gram of radium is

$$\frac{2.92 \times 10^{16}}{3.6 \times 10^{19}} = .0008 \text{ c.c., or } 0.8 \text{ c.mms.}$$

We shall now consider the changes that may be expected to occur in a volume of pure emanation from the point of view of the disintegration theory. The emanation emits α particles and is transformed into the active deposit, which behaves as a type of non-gaseous matter and attaches itself to the walls of the containing vessel. The amount of emanation decreases exponentially, falling to half value in 3.8 days. We should thus expect the volume of the emanation to shrink, and since the activity of the emanation has decayed to a small fraction of its original value after one month, the volume of the emanation after that interval should be very small. The remarkable way in which these theoretical conclusions have been verified will now be considered.

Ramsay and Soddy[1] attacked the difficult problem of isolating the emanation and determining its volume in the following way. The emanation from 60 milligrams of radium bromide in solution was collected for 8 days, and then drawn off through the inverted siphon, E (see Fig. 22), into the explosion burette, F. The radium in the solution produces hydrogen and oxygen at a rapid rate, and the emanation was initially removed with these gases. After explosion, the slight excess of hydrogen mixed with the emanation was left for some time in contact with caustic soda, placed in the upper part of the burette, in order to remove the carbon dioxide present. In the meantime, the upper part of the apparatus had been exhausted as completely as possible. The connection with the mercury pump was then closed, and the hydrogen and emanation allowed to enter the apparatus, passing over a phosphorus pentoxide tube, D, to remove all trace of water vapor.

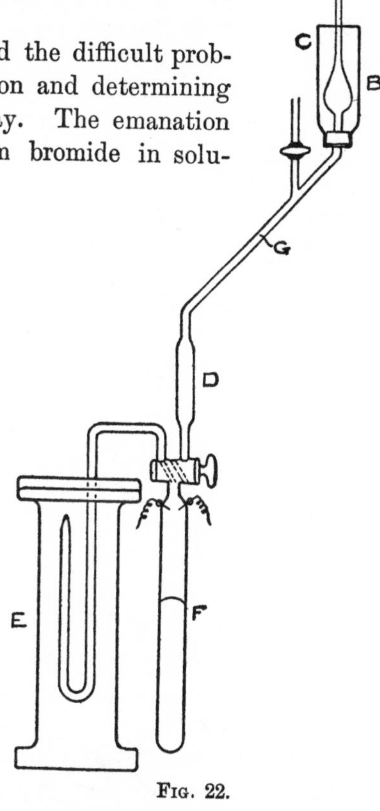

Fig. 22.

Apparatus of Ramsay and Soddy for determining the volume of the radium emanation.

[1] Ramsay and Soddy: Proc. Roy. Soc., lxxiii., p. 346 (1904).

The emanation was condensed in the lower part of the tube B, which was surrounded by liquid air. The process of condensation of the emanation at B was made evident by the brilliant luminosity of the lower part of the tube. The mercury of the burette was allowed to run to A, and the tube AB again completely pumped out. The connection with the pump was again closed, the liquid was removed, and the volatilized emanation forced into the accurately calibrated capillary tube A. Observations were then made for a space of several weeks on the variation in volume of the emanation. The results are shown in the following table:

Time.	Volume.	Time.	Volume.
Start	0.124 c.mms.	7 days	0.0050 c.mms.
1 day	0.027 " "	9	0.0041 " "
3	0.011 " "	11	0.0020 " "
4	0.0095 " "	12	0.0011 " "
6	0.0063 " "		

The volume decreased, and after four weeks only a minute bubble remained, but this retained its luminosity to the last. During this time, the tube was colored a deep purple by the rays. This caused difficulties in readings of the volume, and a strong source of light was found necessary. Ramsay and Soddy consider that the apparent sudden decrease during the first day may have been due to the mercury sticking in the capillary tube. Taking the readings after one day, the volume of the emanation is found to shrink approximately according to an exponential law, decreasing to half value in about 4 days. This is about the rate of decrease of volume to be expected from theoretical considerations. Another experiment was made with a fresh supply of emanation, but a very surprising difference was noted. The gas had an initial volume of 0.0254 c.mm. at atmospheric pressure, and a special series of experiments was made to determine the volume occupied by the gas in the capillary tube at varying pressures. The emanation was found to obey Boyle's law within the limit of experimental error. Unlike its behavior in the first experiment, however, the volume occupied by the gas in the capillary tube, instead of shrinking, steadily increased, and 23 days later was about 10 times the initial value. At the same time, bubbles commenced to appear in the mercury column below the level of the gas.

Further experiments are necessary in order to elucidate the contradictions observed in these two experiments. It will be seen later that the gas helium is a transformation product of the emanation. This appears to have been absorbed in the walls of the tube in the first experiment. Such a result is not unexpected, for there is considerable evidence that the α particles expelled from the radioactive products consist of helium atoms projected with great velocity. Most of these atoms would be buried in the walls of the glass tube to an average depth of about .02 mm., and their diffusion back into the gas may depend on the kind of glass employed. The most plausible explanation is that the helium after absorption by the walls of the glass capillary diffused back into the gas in the second experiment, but not in the first.

Ramsay and Soddy concluded from their experiments that the maximum volume of the emanation released from one gram of radium was slightly greater than one cubic millimetre at standard pressure and temperature.

The theoretical and calculated amounts 0.8 and 1 c. mm., respectively, are thus in very good agreement, and indicate the general correctness of the theory on which the calculations are based.

Spectrum of the Emanation

After the isolation of the emanation, and determination of its volume, a number of experiments were made by Ramsay and Soddy to determine its spectrum. In some of the experiments several apparently new bright lines were seen for a moment, but these rapidly vanished in consequence of the liberation of hydrogen in the tube. Ramsay and Collie[1] continued the experiments, and were finally successful in obtaining the spectrum of the emanation, which lasted for a sufficient interval to determine rapidly the wave-lengths of the more obvious lines by

[1] Ramsay and Collie: Proc. Roy. Soc., lxxiii., p. 470 (1904).

means of eye measurements. The spectrum, however, soon faded, and was finally completely masked by that of hydrogen. They state that the spectrum was very brilliant and consisted of a number of bright lines, the spaces between being perfectly dark. The spectrum bore a striking resemblance in general character to the spectra of the inert gases of the argon family. On repeating the experiment with a fresh supply of emanation, many of the bright lines were seen again, while some new lines, not observed in the first spectrum, made their appearance. They conclude that the emanation undoubtedly has a definite and well-marked spectrum of bright lines.

Heat Emission of the Emanation

One gram of radium in radioactive equilibrium continuously emits heat at the rate of about 100 gram calories per hour. If the emanation is released from the radium by solution or heating, the heating effect of the radium decreases to a minimum of about 25 per cent of the original value, and then as new emanation is formed it gradually increases, reaching its old value after a month's interval. The vessel containing the emanation released from the radium is found to emit heat at a rapid rate, and, three hours after removal, gives out about 75 per cent of the heat emitted by the original radium. The rate of heat emission of the emanation decays at the same rate as it loses its activity, *i. e.*, it falls to half value in about 4 days. The curves of decrease of heating effect of the emanation and of recovery of the heating effect of the radium are, like the activity curves, complementary to each other. The heat emission of the two together is always equal to that of the radium in radioactive equilibrium.

The heat emission of the tube containing the emanation is not due to the emanation alone, but also to the active deposit formed from the emanation. The laws controlling the heat emission of radium and its products will be considered more completely in Chapter X.

It is thus seen that the emanation, together with its transformation products, is responsible for about three quarters of the heat emission of radium. It is difficult to disentangle the heating effect of the emanation from that of its rapidly changing products, but there is no doubt that it supplies about one quarter of the total heating effect of the radium.

Thus one cubic millimetre of the emanation — the maximum amount released from one gram of radium — itself emits heat at the rate of 25 gram calories per hour. Now the heating effect of the emanation falls off at the same rate as its activity. The total heat emission of the emanation during its life is given by Q/λ. The value of λ, with the hour as the unit of time, is $1/132$, and since $Q = 25$, the total heat emitted by the emanation is 3300 gram calories. If we include with that of the emanation the heating effects of its subsequent products, the total heat emitted from the emanation tube is about three times this amount, or 9900 gram calories. This corresponds to a volume of the emanation of about one cubic millimetre. The total heat released from one cubic centimetre of the emanation and its products is thus about ten million gram calories.

Now in the union of hydrogen with oxygen to form water more heat is emitted, weight for weight, than in any other known chemical reaction. In the explosion of 1 c.c. of hydrogen with $\frac{1}{2}$ c.c. of oxygen to form water, 3 gram calories of heat are emitted. We thus see that the transformation of the emanation is accompanied by nearly four million times as much heat as is given out by the union of an equal volume of hydrogen with oxygen to form water.

If we assume that the atom of the emanation has 200 times the mass of the hydrogen atom, it can readily be calculated that one pound weight of the emanation would emit energy at a rate corresponding to 10,000 horsepower. This evolution of energy would fall off exponentially, but, during the life of the emanation, the total energy released would correspond to about 60,000 horsepower-days.

These figures bring out in a striking way the enormous evolution of heat accompanying the changes in the emanation. The amount is of quite a different order of magnitude from that absorbed or released in the most violent chemical reactions.

We shall see later (Chapter X) that probably every radioactive product which expels α particles emits an amount of heat of the same order of magnitude as that emitted by the emanation. In fact, it will be shown that this evolution of heat is a necessary accompaniment of their radioactivity, for the heat is a measure of the kinetic energy of the α particles expelled from the emanation and its products.

Discussion of Results

We may now briefly summarize the properties of the radium emanation discussed in this chapter. (1) The emanation is a heavy gas which does not combine with any substances, but appears to be allied in general properties with the inert group of gases of which helium and argon are the best known examples. (2) It diffuses like a gas of high molecular weight and obeys Boyle's law. (3) It has a definite spectrum of bright lines analogous to the spectra of the inert gases. (4) It is condensed from a mixture of gases at a temperature of –150° C. (5) Unlike ordinary gases, the emanation is not permanent, but undergoes transformation according to an exponential law. The volume of the emanation consequently decreases at the same rate as it suffers disintegration, *i. e.*, its volume shrinks to half value in 3.8 days. The transformation of the emanation is accompanied by the expulsion of α particles, and results in the appearance of a new series of non-gaseous substances deposited on the surface of bodies. The properties of the active deposit, and the changes occurring in it, will be discussed in detail in the next chapter.

The emanation, weight for weight, is about one hundred thousand times as active as the radium from which it is derived. On account of its enormous activity, it glows in the dark and causes a brilliant phosphorescence in many substances. The rays quickly color glass, quartz, and other bodies, and produce a rapid evolution of hydrogen and oxygen in a water solution. The transformation of the emanation is accompanied by an enormous evolution of heat, of an order one million times greater than that observed in any chemical reaction.

We have seen that the emanation and its subsequent products are responsible for three quarters of the activity of radium measured by the α rays. The emanation itself does not emit β or γ rays, but these arise from one of its subsequent products. Consequently the β and γ ray activity of radium is almost completely removed by depriving it of its emanation, provided that several hours have been allowed to elapse in order that the active deposit left behind with the radium may lose its activity.

The emanation, with its subsequent products, thus contains the concentrated essence of the radioactivity of radium. A tube containing the radium emanation has all the radioactive properties of radium in equilibrium. It emits α, β, and γ rays, evolves heat, and produces luminosity in many substances. Radium itself, freed from the emanation and the active deposit, emits only α rays. Its activity and heating effect under such conditions is only one quarter of its usual value, when in radioactive equilibrium.

The emanation is produced from radium at a constant rate, and appears to be a direct disintegration product of the radium atom. Following the same line of argument previously considered, it may be supposed that a minute fraction of the total number of radium atoms explode every second, each violently ejecting an α particle. The radium atom, minus an α particle, becomes the new substance — the emanation. The atoms of the emanation are far more unstable than those of radium itself, and break up with the expulsion of α particles at such a rate that half of the particles are transformed in 3.8 days. After the expulsion of an α particle, the emanation turns into the active deposit.

The transformations, so far considered, and the rays emitted, are graphically illustrated below:

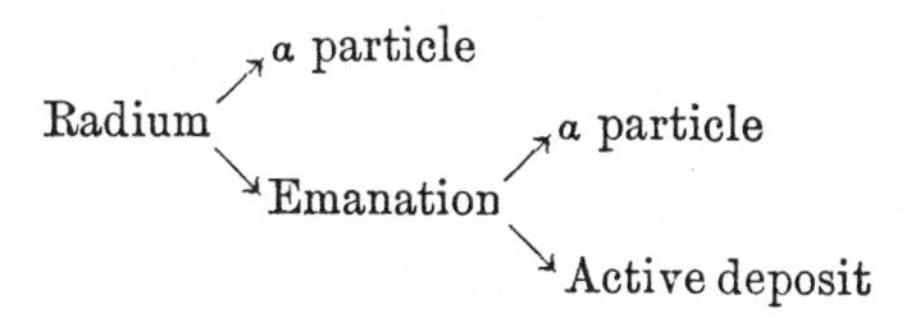

The remarkable differences in the chemical and physical properties of a disintegration product and its parent substance are strikingly illustrated by the comparison of radium with its emanation. Radium is a solid substance of atomic weight 225, closely allied in ordinary chemical properties with barium. It has a definite well-marked spectrum analogous in many respects to the spectra of the rare earths. It is non-volatile at ordinary temperatures, and apart from its radioactivity has all the properties of a new element very analogous to barium. On the other hand, the emanation is an inert gas which cannot be made to combine with any substance. Its spectrum of bright lines is similar in general appearance to the spectra of the helium-argon family of gases. It is condensed at a temperature of $-150°$ C. Apart from its radioactivity, the properties of the emanation are thus entirely different from those of the parent radium, and, if we had no proof of its production by radium, there would be no reason to believe they were in any way connected with each other.

CHAPTER IV

TRANSFORMATION OF THE ACTIVE DEPOSIT OF RADIUM

IN the previous chapter attention has been drawn to the fact that all bodies surrounded by the radium emanation become coated with an invisible active deposit, possessing physical and chemical properties which sharply distinguish it from the emanation. This property of radium of "exciting" or "inducing" activity in neighboring bodies was first observed by P. Curie,[1] and has in recent years been the subject of a number of investigations.

In this chapter, the transformations taking place in this active deposit will be discussed, and it will be shown that, in general, the deposit consists of a mixture of three distinct substances called radium A, B, and C. Radium A arises directly from the transformation of the emanation, radium B arises from radium A, and radium C from radium B.

The three products are thus derived by the successive disintegration of the emanation. The analysis of these stages is somewhat more difficult than in the case of two changes already considered for thorium, but can be attacked by the same general methods.

The active deposit of radium is analogous in many respects to the corresponding deposit produced by the thorium emanation. It is a material substance, which, in the absence of an electric field, is deposited from the gas on the surface of all bodies in contact with the emanation. In a strong electric field it is mostly concentrated on the negative electrode. In this respect it behaves similarly to the active deposit of thorium. The active matter can be partly removed from a platinum wire by solution in hydrochloric acid, and remains behind

[1] M. and Mme. Curie, Comptes rendus, cxxix, p. 714 (1899).

on the dish when the acid is driven off by heat. By using the emanation from about 10 milligrams of radium bromide, a wire can be made intensely active. It causes brilliant fluorescence on a screen of willemite or zinc sulphide brought near it. The deposit is entirely confined to the surface of a conductor. If a strongly active wire is drawn across a screen of willemite or other substance which lights up under the action of the rays, a bright luminous trail is left behind. This is due to the removal of some of the deposit by the particles of the screen over which it has been rubbed. The luminosity left behind gradually decreases, and is very small after 3 hours. The removal of the active deposit by rubbing is also easily shown by bringing near an electroscope a piece of cloth which has been drawn over the active wire. The electroscope is discharged almost instantly, and this discharging property persists, but with diminishing amount, for several hours.

In the case of a short-lived emanation like that of thorium, the excited activity, in the absence of an electric field, is greatest on bodies placed near the emanating thorium compound. This result is to be expected, since the emanation is decomposed before it has time to diffuse far from its source. On the other hand, in a similar enclosure containing radium as a source of emanation, the excited activity is produced on all bodies placed in the vessel. In this case the life of the emanation is long compared with the time taken for the emanation to be distributed by the processes of diffusion to all parts of the enclosure.

Bodies which are completely screened from the direct radiation of the radium become active. This is clearly brought out in an experiment made by P. Curie, which is shown in Fig. 23.

A small open vessel, a, contained a radium solution, giving off emanation at a constant rate. This was placed in a closed vessel in which plates A, B, C, D, E were fixed in various positions. After a day's exposure, all the plates on removal were found to be active, even that in the position D, completely shielded from the direct radiation of the radium by a lead block P.

The amount of activity per unit area on a plate in a given position is independent of the material of the plate. A plate of mica becomes just as active as one of metal. The amount of excited activity on a given area depends to some extent on the free space in the neighborhood. The lower surface of the plate A, for example, would be less active than the upper surface, since the active deposit on the lower side arises mainly from the small volume of emanation between it and the enclosure, while the upper surface of the plate gains the active deposit generated in a much larger volume.

The emanation from several milligrams of radium bromide causes so great an activity on a wire or metal plate exposed in its presence, that the ionization current produced by it can be readily measured by a sensitive galvanometer. With such intensely active plates, a large voltage is required to produce a saturation current through the gas unless the plates of the testing vessel are placed close together.

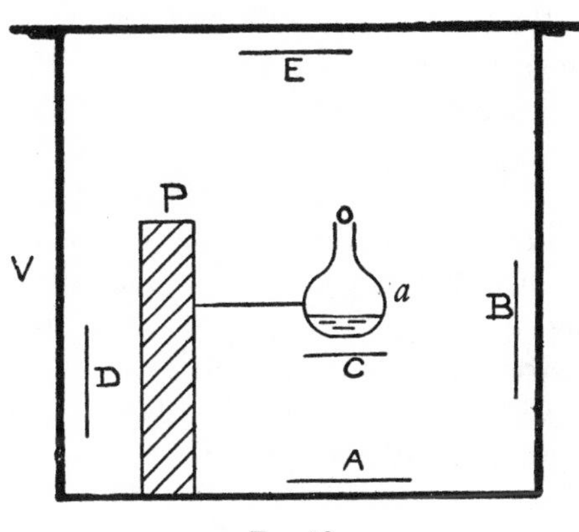

FIG. 23.

Distribution of excited activity on bodies in the presence of the radium emanations.

We shall first consider the evidence in support of the view that the active deposit is a disintegration product of the radium emanation. If some radium emanation is introduced into a cylindrical testing vessel such as is shown in Fig. 10, and the ends closed, the activity, measured by the saturation current through the gas, increases with the time for several hours, generally reaching about twice the value observed at the moment of introduction of the emanation. The comparative increase, however, varies to some extent with the dimensions of the testing vessel, on account of the difference in penetrating power of the a rays emitted by the various products.

When the emanation is blown out, the active deposit is left behind, and loses the greater part of its activity in a few hours. This property of producing an active deposit is not shown by radium which has been freed from emanation, but belongs to the emanation alone. The excited activity produced in bodies is directly proportional to the amount of emanation present, no matter how old the emanation may be. For example, if the emanation, which still remains after being stored in a gas holder for a month, is passed into a testing vessel, excited activity is still produced, and in an amount which bears the same ratio to the activity of the emanation present as for a new sample of emanation tested immediately after its release from radium.

The constancy of this ratio between the amount of the emanation present and the amount of active deposit produced is at once explained if the emanation is the parent of the active deposit. For example, suppose that a body is exposed to a constant supply of emanation. The activity imparted to the body reaches a steady limit after about 5 hours. There is, then, a state of equilibrium between the active deposit and the emanation. Under such conditions, the number of atoms of radium A which break up per second must equal the number of new atoms of radium A supplied per second by the decomposition of the emanation. This in turn is equal to the number of atoms of emanation which break up per second. A similar result also holds true for radium B and C. Since the number of atoms of any individual product which break up per second is always proportional to the total number present, it is seen that the equilibrium number of atoms of radium A must always be proportional to the number of atoms of emanation. If λ is the constant of decay of the emanation and λ_A, λ_B, λ_C, the constants for radium A, B, and C, respectively, then the equilibrium amounts N_A, N_B, N_C, respectively, of the three products are given by the equations

$$\lambda_A N_A = \lambda_B N_B = \lambda_C N_C = \lambda N,$$

where N is the total number of atoms of emanation present. When a state of equilibrium has been reached, the number of atoms of each product present will be different, being directly proportional to the period of each product. A rapidly changing substance will consequently be present in less amount than a slowly changing one.

After introducing the emanation into a closed vessel, its amount, as we have seen, decreases exponentially. Since, however, the periods of the products of the active deposit are small compared with that of the emanation itself, the amount of the active deposit will, after a few hours, nearly reach an equilibrium value, and will then decrease *pari passu* with the emanation.

The excited activity will thus fall off at the same rate as the activity of the emanation. This proportion has been utilized, as we have already seen, by Curie and Danne, to determine the constant of decay of the emanation by measurement of the β and γ rays which escape from the active deposit through the walls of a closed vessel containing the emanation.

Activity Curves of the Active Deposit

We shall now consider in detail the variation with the time of the activity of this deposit under different conditions. The experimental results are at first sight very complicated, for the activity curves not only vary remarkably with the time of exposure to the emanation, but also depend on whether the α, β, or γ rays are used as a means of measurement. It is thus very necessary in each case to specify carefully, not only the time of exposure to the emanation, but also the type of rays used for measurement.

The decay curves of the active deposit are independent of the nature and size of the body that has been made active and of the amount of emanation to which it has been exposed. If a wire is to be made active the arrangement shown in Fig. 24 is very suitable.

A solution of radium is placed in a vessel closed by a rubber stopper. The emanation collects in the air space above the solution. The thin wire, W, to be made active is fixed into a fine hole bored in the end of a central rod. This rod slips freely through an ebonite cork fixed in a brass tube, B. A platinum wire P passes through the rubber cork and dips into the solution.

The platinum wire is in metallic connection with the brass tube. The central rod is connected with the negative pole of a battery of 300 or 400 volts, and the platinum wire with the positive. Under such conditions, the moist walls of the glass vessel, the solution, and the tube B, are charged positively, and the wire W is the only negatively charged body in the presence of the emanation. The active deposit is consequently concentrated upon it, and, in the presence of a large amount of emanation, the activity of the wire becomes very great.

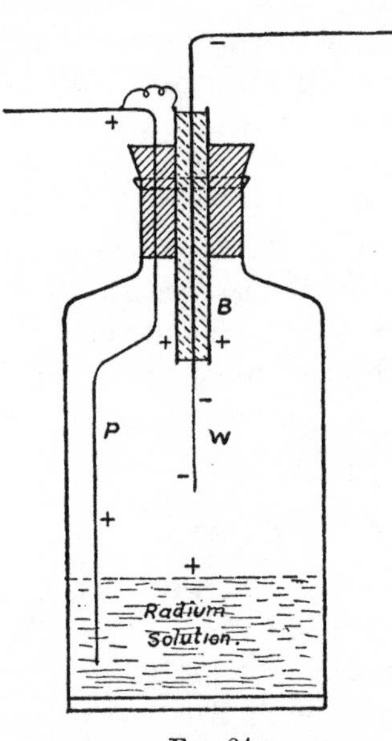

Fig. 24.

Arrangement for concentrating the active deposit derived from the radium emanation on a small negatively charged wire.

After introducing the wire, a little hard wax is run round the top of the rod, to prevent the escape of the emanation. When the wire has been exposed for the interval required, the rod is removed and the active wire released. Since the fine wire is of smaller diameter than the rod, the wire need not touch the side during removal, so that none of the active deposit is rubbed off.

In order to test the variation of the a ray activity of this wire with time, it is attached to the end of a brass rod forming the central electrode of a testing vessel such as is shown in Fig. 10.

If a greater surface is to be made active, a sheet of metal is placed in a glass tube closed at both ends. The emanation is introduced after first exhausting the vessel, and the active matter is then deposited upon the metal by the process of diffusion. After removal, the activity of the plate is tested electrically, using a parallel plate apparatus similar to that described in Fig. 9.

a Ray Curves

We shall first consider the decay of activity, measured by the a rays, for a body exposed for a short time in the presence of the emanation. The time of exposure — not more than one

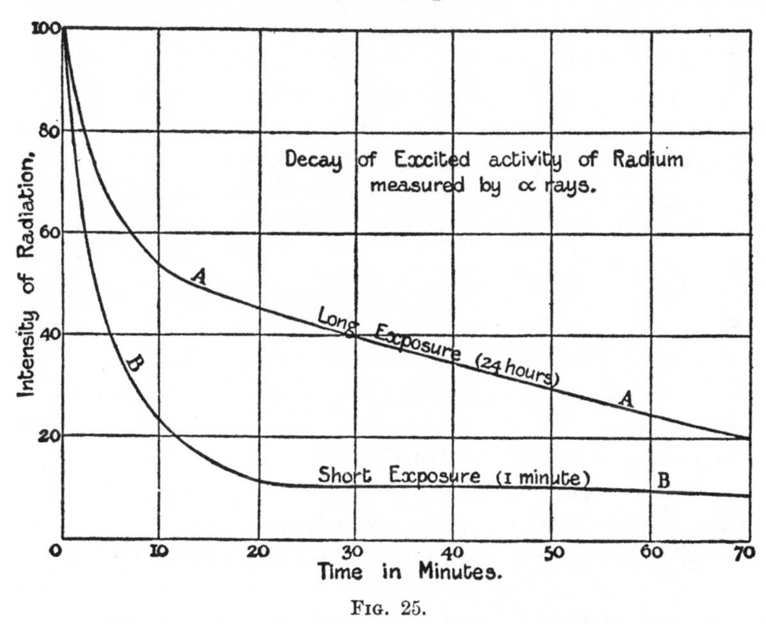

Fig. 25.

minute — is supposed to be short compared with the period of the changes in the active matter. The results are shown in Fig. 25, Curve BB, the maximum activity immediately after removal being taken as 100.

The activity at first decreases very nearly according to an exponential law, falling off to half value in about 3 minutes. After 20 minutes, the activity is less than 10 per cent of the initial value, and remains nearly constant for a further 20 minutes, and then gradually decays. After several hours the activity again decreases nearly exponentially with a period of 28 minutes.

In the same figure (Curve AA) is shown the a ray decay curve for a long exposure. The time of exposure in this case (about 5 hours will suffice) is supposed to be sufficient to allow the active deposit and the emanation to have very nearly reached a stage of radioactive equilibrium. There is initially a rapid decay with a 3 minute period, and then a gradual decrease at a slower rate than is given by an exponential law. After about 5 hours the decay curve is nearly exponential, falling to half value in about 28 minutes.

The initial rapid change with a 3 minute period is due to the product radium A. The final exponential decay with a 28 minute period shows that another product, having a 28 minute period, is also present. Before discussing the explanation of the intermediate portion of the two curves the activity curves measured by the β and γ rays will first be considered.

β Ray Curves

In order to determine the β ray curves, an electroscope was used. The active plate or wire was placed under the base of the electroscope, which was covered with a sheet of aluminium of sufficient thickness to absorb all the a rays. The discharge produced in the electroscope is then due to the β and γ rays together, the effect of the former preponderating. The curve in Fig. 26 shows the variation of the β ray activity with time, for a wire which had been exposed for one minute in the presence of a large amount of emanation. It will at once be observed that the curve is entirely different in character from the corresponding a ray curve shown in Fig. 25. The β ray activity is small at first, but increases with time, reaching a maximum after about 35 minutes. Several hours later it decays nearly exponentially with a period of 28 minutes.

The β ray curve for a long exposure to the emanation is shown in Fig. 27.

The curve is very different in shape from the short exposure curve. The activity does not increase initially, but falls, first slowly and then more quickly. Finally, as in the other cases, it decreases exponentially with a period of 28 minutes.

γ Ray Curves

The curves for a short and long exposure measured by the γ rays alone are identical with those obtained for the β and γ rays together. The measurements were made with an electroscope, the rays passing through about 1 cm. of lead before

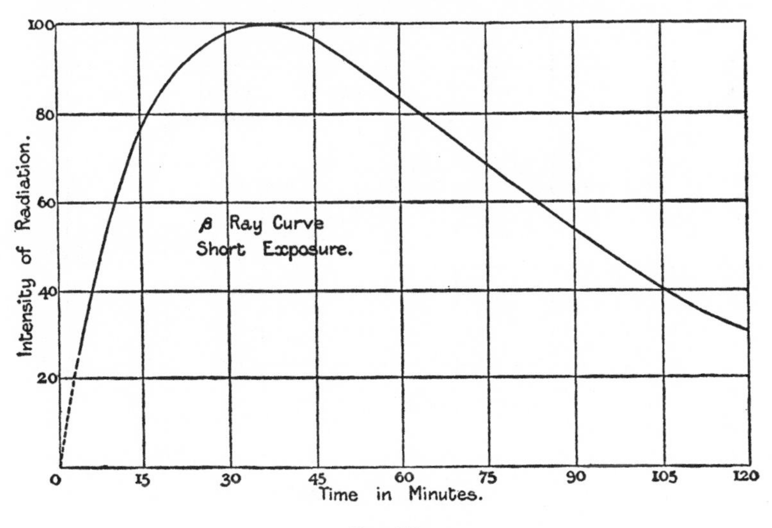

Fig. 26.

Variation of the activity, measured by the β rays, of a body exposed for a short interval to the radium emanation.

entering the electroscope. This insures that the β as well as the α rays are cut off completely.

The identity of the β and γ ray curves shows that the two kinds of rays always occur in the same proportion. This relation is a strong argument in favor of the view that the γ rays are a type of X-rays, which are set up at the moment of the expulsion of the β particle from radioactive matter. This ratio between the intensities of the two kinds of rays has been shown to hold in every case so far examined, and suggests that the

γ rays bear the same relation to the β rays that the X-rays bear to the cathode rays.

THEORY OF SUCCESSIVE CHANGES IN RADIUM

We shall show later that the peculiarities of the decay curves of the active deposit of radium for any time of exposure, whether

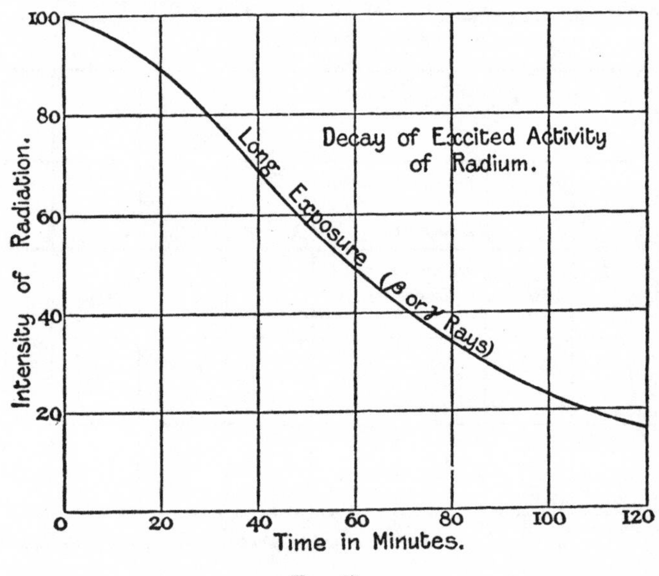

FIG. 27.

Variation of the activity, measured by the β or γ rays, of a body exposed for a long interval to the radium emanation.

the activity is measured by the a, β, or γ rays, can be satisfactorily explained on the following assumptions: —

(1) That the emanation is transformed into a product called radium A, which emits only a rays, and has a period of 3 minutes.

(2) That radium A is transformed into radium B, which has a period of 28 minutes, and is transformed without the emission of a, β, or γ rays. In other words, radium B is a rayless product.

(3) Radium B is transformed into radium C, which has a period of 21 minutes, and emits during its transformation α, β, and γ rays.

We thus have to deal with the problem of three successive changes. Since, however, the first product, radium A, is rapidly transformed with a 3 minute period, the amount of it remaining, for example, 21 minutes after removal, is only 1/128 of the initial amount.

For simplicity, therefore, in the discussion of the activity curves measured by the β rays, we shall for the moment disregard the first rapid change, and suppose that the emanation is transformed directly into radium B. As a matter of fact, it is found that the experiments agree better with theory if the first transformation is disregarded altogether. A possible explanation of this peculiarity in the curves will be considered later.

In the discussion of the activity curves for the active deposit of thorium, it has been shown that the experimental curve for a short exposure may be satisfactorily explained if the emanation is supposed to change into the rayless product, thorium A, which has a period of 11 hours. This in turn is transformed into thorium B, which emits α, β, and γ rays, and has a period of about 1 hour. These results deduced from analysis of the activity curves have been completely substantiated by experiments in which the products thorium A and B have been separated from each other by various physical and chemical methods.

The case of radium is very analogous, for, disregarding the first 3 minute change, the product radium B emits no rays, but changes into radium C, which emits α, β, and γ rays.

We shall now consider the theory of two successive changes of the character explained above.

Let λ_1, λ_2 be the constants of change of the products radium B and C respectively.

Let P and Q be the number of atoms of B and C respectively present at any time after removal from the emanation.

Two general cases will first be considered, corresponding to a short and long exposure of a body to the radium emanation.

Case of a Short Exposure

The matter initially deposited is supposed to be all of one kind, radium B. Let n be the number of particles of B that have been deposited. The number, P, of these remaining at any time t, after removal, is given by

$$P = n\,e^{-\lambda_1 t}.$$

We have shown on page 50 that the rate of change of the number Q of atoms of C existing at any time, t, after removal, is given by

$$\begin{aligned}\frac{dQ}{dt} &= \lambda_1 P - \lambda_2 Q \\ &= \lambda_1 n\,e^{-\lambda_1 t} - \lambda_2 Q.\end{aligned} \tag{1}$$

The solution of this equation (see page 51) shows that Q is given by

$$Q = \frac{n\,\lambda_1}{\lambda_1 - \lambda_2}\,(e^{-\lambda_2 t} - e^{-\lambda_1 t}).$$

The number of atoms of P and Q existing at any time after removal is shown in Fig. 28. The initial number of atoms of B deposited is supposed to be 100. The exponential curve, BB, expresses the amount of B remaining unchanged at any time. The curve CC shows the number of atoms of radium C existing at any time. The periods of the changes of B and C are about 28 and 21 minutes respectively, so that

$$\lambda_1 = 4.13 \times 10^{-4}\,(\text{sec.})^{-1}, \qquad \lambda_2 = 5.38 \times 10^{-4}\,(\text{sec.})^{-1}.$$

The amount of radium C, initially zero, increases to a maximum in about 35 minutes, and then diminishes, and about 5 hours later decays exponentially, with a period of 28 minutes. The amount of C will thus decrease, not according to its own period, but according to the longer period of the rayless product. This is easily shown from the equation for Q, which may be expressed in the form

$$Q = \frac{n\,\lambda_1\,e^{-\lambda_1 t}}{\lambda_2 - \lambda_1}\,(1 - e^{-\overline{\lambda_2 - \lambda_1}t}),$$

After 7 hours, $\qquad e^{-(\lambda_2 - \lambda_1)t} = .043,$

and is thus almost negligible. Q then varies very nearly as $e^{-\lambda_1 t}$, *i. e.*, according to the period of the rayless product.

Since B does not emit rays and C does, the value of Q at any time is proportional to the activity of the mixture of products B and C.

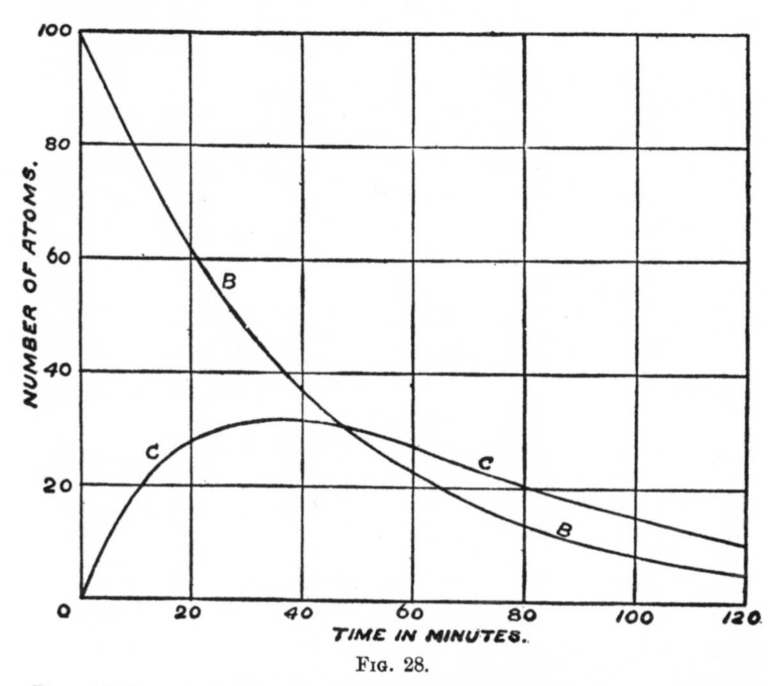

FIG. 28.

Theoretical curves showing the variation of the number of atoms of radium B and radium C when the matter present initially consists only of radium B.

The curve CC should thus be identical in form with the curves for a short exposure measured by the β or γ rays, and within the limits of experimental error this is found to be so.

CASE OF A LONG EXPOSURE

Suppose that P_0 and Q_0 are the equilibrium numbers of atoms of B and C present after a long exposure to the emanation.

Under such conditions

$$\lambda_1 P_0 = \lambda_2 Q_0 = q,$$

where q is the number of atoms of emanation breaking up per second.

The value of P, the number of atoms of radium B present at any time t after removal from the emanation is given by

$$P = P_0 e^{-\lambda_1 t} = \frac{q}{\lambda_1} e^{-\lambda_1 t}.$$

The value of Q is given by equation (1) as before. The solution of this equation is of the form

$$Q = a e^{-\lambda_1 t} + b e^{-\lambda_2 t}.$$

By substitution in equation (1) it is seen that

$$a = \frac{q}{\lambda_2 - \lambda_1}.$$

Since initially when $t = 0$, $Q = Q_0 = \frac{q}{\lambda_2}$,

we have $$a + b = \frac{q}{\lambda_2};$$

thus $$b = \frac{-q\lambda_1}{\lambda_2(\lambda_2 - \lambda_1)},$$

and $$Q = \frac{q}{\lambda_2 - \lambda_1}\left(e^{-\lambda_1 t} - \frac{\lambda_1}{\lambda_2} e^{-\lambda_2 t}\right). \qquad (2)$$

The variation of the amount of radium B with time after a long exposure is shown in Fig. 29, the number of atoms of B initially present being 100. The number of atoms of radium C present initially is $\frac{\lambda_1}{\lambda_2} P_0$.

The curve, CC, expressing the number of atoms of C present at any time thus begins at a point whose ordinate is 77 instead of 100.

Since the β or γ ray activity of C is proportional at any time to the value of Q, the curve showing the variation of radium C with time should be of the same form as the activity curve in Fig. 27 for a long exposure, as measured by the β and γ rays. This is the case, for the theoretical and observed curves agree within the limit of experimental error. This is shown in the following table:

Decay of Activity Measured by the β Rays for a Long Exposure to the Emanation.

Time in minutes after removal from emanation.	Observed activity.	Theoretical activity.
0	100	100
10	97.0	96.8
20	88.5	89.4
30	77.5	78.6
40	67.5	69.2
50	57.0	59.9
60	48.2	49.2
80	33.5	34.2
100	22.5	22.7
120	14.5	14.9

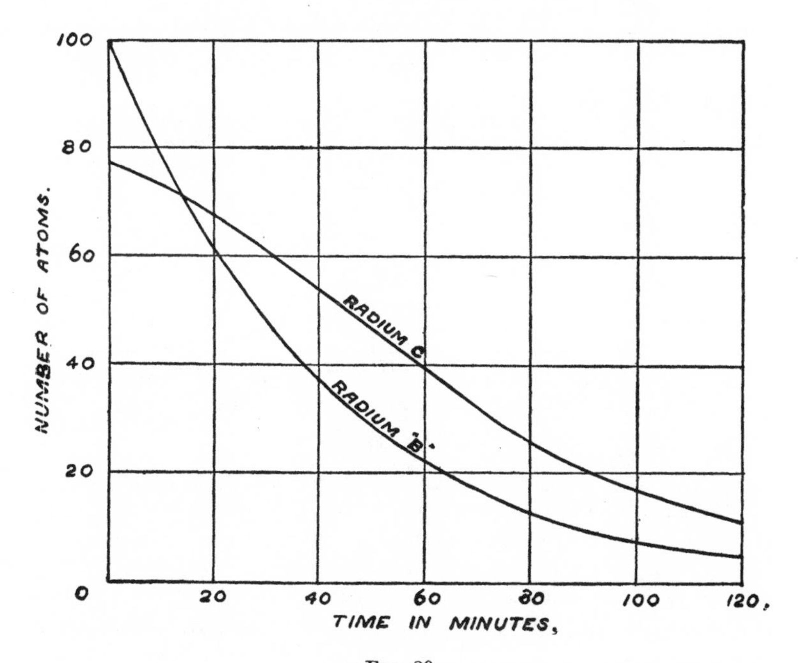

Fig. 29.

Theoretical curves showing the number of atoms of radium B and radium C existing at any time, when the matter initially consists of radium B and C in radioactive equilibrium.

The fact that the long exposure curve shown in Fig. 27 results from two successive products, the first of which does not emit rays at all, can readily be shown by graphical analysis.

Immediately after removal of the active body from the emanation, the active deposit consists of B and C in equilibrium. The β ray activity observed is due entirely to C, and, leaving

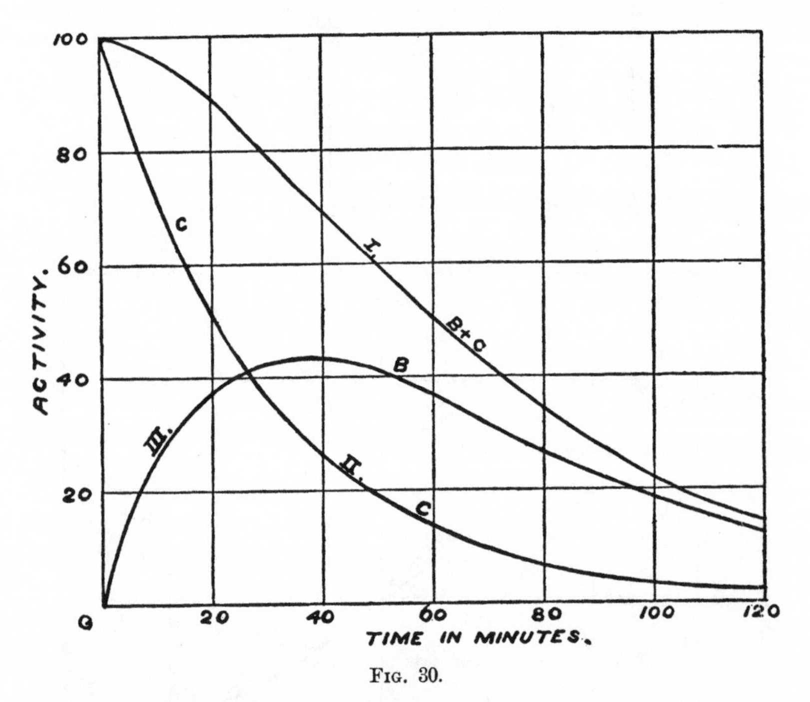

Fig. 30.

Analysis of the β ray curve for a long exposure to the emanation, in order to show that it results from the presence of two products, the first of which is rayless.

out of account for the moment the fresh supply of C from the disintegration of B, the amount of C must, if left to itself, diminish exponentially following the period of C, *i. e.*, the activity will fall to half value in 21 minutes. This decay curve CC is shown in Fig. 30. Now the difference at any time between the ordinates of the observed curve B + C, and the theoretical curve, CC, must be due to the activity of C, supplied by the breaking up of B. This difference curve, BB (see Fig. 30), should be identical in shape with the β ray curve of the active deposit for a short exposure. This must evidently be the case, since this curve gives the activity arising from the transformation of B alone, all the matter present initially being of the kind B, which changes into C.

By comparison of the curve, BB, with the short exposure curve shown in Fig. 27, this identity is seen to hold. The activity rises from zero, and reaches a maximum after 35 minutes and then decays.

It is of interest to observe that the empirical equation of the decay curve of the β ray activity for a long exposure to the emanation was obtained before the theoretical explanation was advanced. Curie and Danne[1] found that the activity I_t at any time could be expressed by an equation of the form

$$\frac{I_t}{I_0} = a e^{-\lambda_1 t} - (a-1) e^{-\lambda_2 t},$$

where $\lambda_1 = 4.13 \times 10^{-4} (\text{sec})^{-1}$ and $\lambda_2 = 5.38 \times 10^{-4} (\text{sec})^{-1}$,

and $a = 4.20$ is a numerical constant. The constant λ_1 was determined from the observed fact that the activity several hours after removal from the emanation decayed exponentially with a period of 28 minutes. The values of a and λ_2 were determined so as to fit the curve. Now this equation is identical in form with the theoretical equation for the activity when the first change is rayless with a period of 28 minutes, and the second change, which has a period of 21 minutes, gives out rays. This is easily seen to be the case. From equation (2), the amount Q of radium C, existing at any time t, is given by

$$Q = \frac{q}{\lambda_2 - \lambda_1} \left(e^{-\lambda_1 t} - \frac{\lambda_1}{\lambda_2} e^{-\lambda_2 t}\right).$$

Initially $$Q = Q_0 = \lambda_2 q.$$

[1] Curie and Danne: Comptes rendus, cxxxvi, p. 364 (1903).

Since the activity at any time is proportional to the amount of C present, *i. e.*, to the value of Q,

$$\frac{I_t}{I_0} = \frac{Q}{Q_0} = \frac{\lambda_2}{\lambda_2 - \lambda_1} e^{-\lambda_1 t} - \frac{\lambda_1}{\lambda_2 - \lambda_1} e^{-\lambda_2 t}.$$

On substituting the values of λ_1, λ_2, which correspond to periods of 28 and 21 minutes respectively,

$$\frac{\lambda_2}{\lambda_2 - \lambda_1} = 4.3 \quad \text{and} \quad \frac{\lambda_1}{\lambda_2 - \lambda_1} = 3.3.$$

Thus the theoretical equation not only agrees in form with that deduced from observation, but the values of the constants are very concordant.

Such a relation between theory and experiment would be widely departed from if B as well as C gave out β rays.

Analysis of the α Ray Curves for a Long Exposure

We are now in a position to analyze the α ray activity curve for a long exposure into its three components. In this case we must take into account the first product, radium A, which emits α rays. The observed α ray curve is shown in Fig. 31, curve A + B + C. This curve was obtained by means of a galvanometer. A piece of platinum foil was placed for several days in a glass vessel containing a large supply of radium emanation. The foil was then rapidly removed, placed on the lower plate of a testing vessel, and a saturating voltage applied. The variation of the α ray activity was measured by means of a high resistance galvanometer. The initial value of the current at the instant of removal was deduced by continuing the curve backwards to meet the vertical axis.

The variation of activity with time is shown in the following table:

Time in minutes.	Activity.	Time in minutes.	Activity.
0	100	30	40.4
2	80	40	35.6
4	69.5	50	30.4
6	62.4	60	25.4
8	57.6	80	17.4
10	52.0	100	11.6
15	48.4	120	7.6
20	45.4		

The activity due to A alone has almost vanished after 20 minutes. At the 20 minutes point the curve B + C is produced backwards, meeting the axis at L. It cuts it at about 50. The difference between the ordinates of the curves A + B + C and LL represents the activity due to radium A, and is shown by

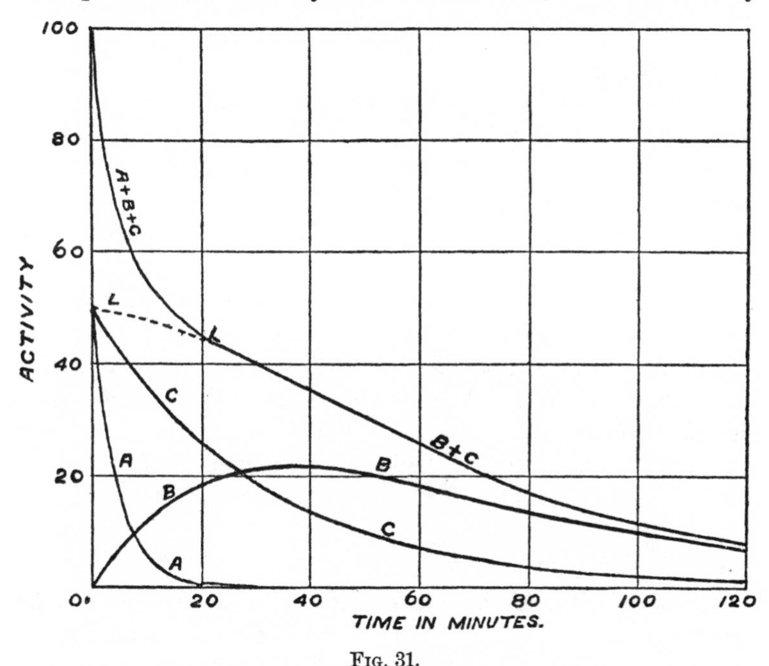

Fig. 31.

Analysis of the α ray curve for a long exposure to the radium emanation, showing that it results from the presence of three products, A, B, and C. Radium A and C emit α rays, while radium B does not.

the curve AA, in the figure. The curve A is exponential with a period of 3 minutes. The curve LL, B + C, is identical in form over its whole range with the activity curve, for a long exposure (see Fig. 27), measured by the β rays. We may conclude from this result that radium B emits no α rays. We know already that it emits no β rays; hence radium B must be a rayless product.

The curve LL, B + C, can be analyzed into its two components exactly in the same way as the corresponding β ray curve. The curve CC represents the variation in the activity of the radium C which existed at the moment of removal from the emanation. The curve BB represents the activity due to C, supplied by the change of B into C. This curve, BB, is identical in form with the β ray curve for a short exposure (see Fig. 26).

We may conclude from this analysis that the active deposit consists of three products, radium A, B, and C, which have the following peculiarities: —

Radium A emits only α rays, and is half transformed in 3 minutes.

Radium B is a rayless product, and is half transformed in 28 minutes.

Radium C emits α, β, and γ rays, and is half transformed in 21 minutes.

Several hours after removal from the emanation, the activity, whether for a long or short exposure, and whether measured by the α, β, or γ rays, falls off exponentially, with a period of 28 minutes. This is due to the fact that the longer period of the rayless product, B, governs the final rate of decay, although the activity is really supplied by the product C of 21 minute period.

Are Radium A and B Successive Products?

For simplicity, in the above comparison of theory with experiment, the effect of radium A on the subsequent changes has been neglected. If radium B is derived from radium A, the amount of A present, when in radioactive equilibrium with the emanation, is 3/28 or .11 of radium B. If A changes into

B according to a 3 minute period, the greater part of A will be transformed in about 15 minutes, and it can be deduced that the amount of B present after that interval should be about 8 per cent greater than if A did not change into B. The effect of this on the subsequent decay curves should be easily measurable under suitable conditions. This point has been examined by the writer,[1] but it was found that theory and experiment agreed much more accurately if A and B were considered to be independent products, separately produced during the transformation of the emanation. The examination of the a ray curve for a short exposure to the emanation does not give definite evidence in either direction.

The conclusion that A and B are independent products, however, involves such important theoretical consequences that before accepting it, a close examination must be made to see if the theoretical conditions are completely realized in practice.

The theory assumes that radium A should be deposited on the electrode very shortly after its production, and that neither A nor its subsequent products escape from the electrode; or, in other words, that these products show no appreciable volatility at ordinary temperatures.

There is no doubt, however, that, under ordinary conditions, appreciable quantities of both radium A and B, and sometimes C, are present, mixed with the emanation, showing that all of these products do not rapidly diffuse to the electrodes. In addition, Miss Brooks[2] has shown that radium B is undoubtedly volatile at ordinary temperatures.

Experiments are at present in progress in the laboratory of the writer to decide whether such divergences between the theoretical and experimental conditions are sufficient to account for the decay curves on the assumption that A and B are successive products. The question is still *sub judice*, but it is hoped that a definite answer will soon be forthcoming.[3]

[1] Rutherford: Bakerian Lecture, Phil. Trans. A, p. 169, 1904.

[2] Miss Brooks: Nature, July 21, 1904.

[3] The cause of this discrepancy between theory and experiment has been indicated by some recent experiments of H. W. Schmidt (Physik. Zeit., 6, No. 25, p. 897, 1905).

If radium A and B are proved to be independent, it will be necessary to suppose that the emanation breaks up into two distinct products, and also expels one or more a particles. The observed fact that the activity due to radium A for a long exposure is nearly equal to that of radium C when the plates are close together, is in agreement with both hypotheses, provided that it is assumed in the non-successive hypothesis that each atom of the emanation breaks into two products, besides expelling an a particle.

On such a view, the emanation gives rise to two distinct families of products. While a possible change of this character is of interest and importance, evidence of an undeniable character must be forthcoming before it can be accepted. If a product can be separated from radium or its active deposit, which decays exponentially, with a 3 minute period, and does not give rise to radium B and C, the independence of A and B will be completely established.

Effect of Temperature on the Active Deposit

It has been assumed without proof in the above discussion that radium B rather than radium C has the period of 28 minutes. The comparison of theory with experiment does not throw any light on the question, since the activity curves are the same if the periods of the products B and C are interchanged.

As in the case of thorium, it is necessary to have recourse to other evidence to settle whether the period of 28 minutes belongs to B or C. It is necessary by some physical or chemical means to isolate B from C, and to examine separately their rates of change.

This has been effected by taking advantage of the greater

He finds that radium B is not a rayless product, but emits β rays which are of much smaller penetrating power than those from radium C. We have seen (Fig. 26) that the β ray curve for a short exposure to the emanation passes through a maximum 35 minutes after removal. This only holds when the rays have been passed through a screen of sufficient thickness to absorb the β rays from radium B. With thinner screens, the maximum is reached earlier. When this new factor is fully taken into account, it appears probable that the experimental curves will completely agree with the theory that radium A, B, and C are successive products.

volatility of radium B when an active wire is exposed to a high temperature. Miss Gates [1] observed that the active deposit of radium was volatilized at a white heat, and redeposited on the cold bodies in the neighborhood. Curie and Danne [2] examined this effect in more detail, and obtained some very interesting results. An active wire surrounded by a cold metal cylinder was heated for a short time by an electric current, and the activity both of the wire itself and of the interior of the cylinder separately examined. At about 400°C., some of the radium B was volatilized. This was established by noting the variation of the activity of the distilled part after the heating. This activity was small at first, passed through a maximum, and then decayed, in exactly the same way as the β ray activity of a body for a short exposure (see Fig. 26). This showed that the matter initially deposited on the cylinder consisted only of the rayless product B, which changed into the ray product C. At a temperature of about 600°C. most of the B was driven off, and also some C.

A number of experiments were then made on the decay of activity of the wire for temperatures varying between 15°C. and 1350°C. At a temperature of 630°C., they state that the activity of the wire decreased exponentially with a period of 28 minutes. The period steadily decreased to 20 minutes, while the temperature rose to 1100°C. At this temperature it passed through a minimum value, and then increased to 25 minutes at 1300°C.

Since their decay curves were exponential, Curie and Danne supposed that all of B was volatilized at 630°C. If this were the case, the results indicated that the 28 minute period must be ascribed to radium C and the 21 minute period to radium B. The experiments also indicated that rise of temperature above 630°C. to 1100°C. produced an apparent alteration in the rate of change of radium C. This, if correct, was a most important result, for there was no previous evidence that temperature had any effect in altering the rate of transformation of a radioactive

[1] Miss Gates: Phys. Rev., p. 300, 1903.

[2] Curie and Danne: Comptes rendus, cxxxviii, p. 748, 1904.

product. According to their experiments, the rate of transformation of radium C was altered in an unexpected manner by rise of temperature, for it increased up to 1100°C. and at a still higher temperature fell again nearly to its normal value.

A close examination of the effect of temperature on the active deposit was recently made by Dr. Bronson[1] in the laboratory of the writer. The results obtained by him showed conclusively that a rise of temperature to 1100° C. has no effect in altering the rate of transformation of the active deposit, and that the experimental results obtained by Curie and Danne can be explained by supposing that in most of their experiments the deposit on the wire after heating consisted not entirely of C but of a mixture of B and C.

In order to test definitely whether temperature has any effect on the decay of the active deposit, an active copper wire was placed in a small length of combustion tubing, and the glass sealed under diminished pressure. This was then heated in an electric furnace to different temperatures. The glass was found to withstand a temperature of about 1100°C. The β ray activity was then carefully examined over a long interval. Between 2.5 and 4 hours, the curves obtained were approximately exponential with a period of 28 minutes. After 6 hours, the curve was accurately exponential with a period of about 26 minutes. Within the limit of experimental error, the curves of decay up to 1100°C were found to be identical with the normal decay curves at atmospheric temperature.

In this experiment, none of the distilled products were able to escape, so that we may conclude with certainty that a rise of temperature up to 1100°C. has no appreciable effect on the rate of transformation of the active products.

On repeating the experiments of Curie and Danne, Bronson found that the decay of activity after heating the wire to a constant temperature was very variable, and that approximately exponential curves were obtained with periods lying between 25 and 19 minutes after heating the wire to the same temperature. If, for example, a current of air was blown through the

[1] Bronson: Amer. Journ. Sci., July, 1905.

electric furnace before removing the wire, the activity fell exponentially with a period of about 19 minutes. In a similar way if a cold copper wire was introduced above the active wire the period had about the same value. In such cases, there is a better opportunity for the distilled part, radium B, to escape from the wire. Several curves were obtained which were accurately exponential with a period of 19 minutes. This result showed that the active product, radium C, has a period of 19 minutes, and that the 26 minute period must be ascribed to radium B.

It was observed that in all cases in which the activity decayed initially with a period lying between 19 and 26 minutes, the curves were not at first accurately exponential. The period always tended towards a value of 26 minutes as the activity decayed, and the law was then exponential. This is exactly what is to be expected if the activity after heating the wire results from a mixture of B and C, the amount of C initially predominating. The activity will first decay to half value with a period intermediate between those of B and C. The amount of radium B, which changes more slowly than C, after a time begins to predominate, and ultimately governs the final rate of decay, *i. e.*, the activity falls finally according to an exponential law with a period of 26 minutes.

The experiments have thus shown that while B is more volatile than C, in many cases all of the B is not removed even if the wire is heated to a temperature far above its point of volatilization.

The periods of the two products, radium B and C, are 26 and 19 minutes respectively, values somewhat lower than the periods 28 and 21 minutes assumed in the previous calculations. Between 2 and 4 hours after removal from the emanation, the activity under normal conditions falls approximately exponentially with a period of 28 minutes, and this originally led to the choice of 28 minutes as the value of one of the periods. The decay curve, however, is not accurately exponential until about 6 hours after removal from the emanation, and the period is then 26 minutes.

In the analysis of the changes, radium C has been given the

shorter period, but the original determinations of the periods of B and C viz., 28 and 21 minutes, have been retained. Over the range considered, the theoretical curves for periods of B and C of 26 and 19 minutes respectively do not differ much from those with the periods of 28 and 21 minutes.

The retention of the old values brings out more clearly the methods originally employed of proving that B was a rayless product and that C gives out α, β, and γ rays. A more accurate determination of the various curves of decay during the first two hours is at present in progress.

The series of transformation products of radium which have so far been discussed are diagrammatically shown in Fig. 32.

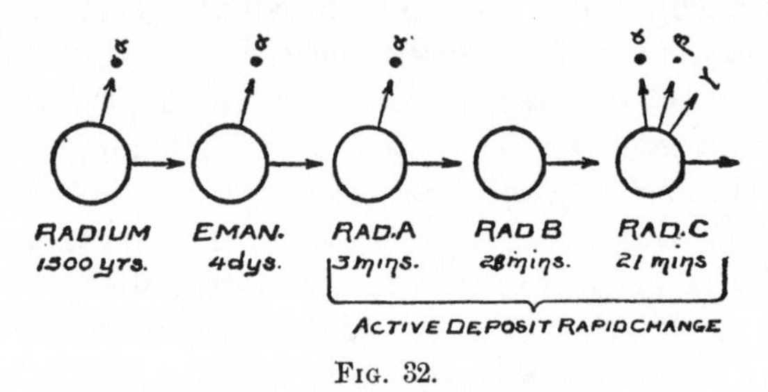

FIG. 32.

Radium and its family of rapidly changing products.

The periods of the products are given for convenience, as well as the character of the emitted rays.

It is a matter of remark that of these five radioactive substances, only radium C gives out β and γ rays. The others emit only α rays. The emission of the α rays is, however, accompanied by a secondary radiation, which is produced probably by the impact of the α rays on matter and consists of electrons which are projected at a speed small compared with that of the β rays proper, and which are consequently very easily deflected by a magnetic field. The presence of such slow moving electrons was first observed by J. J. Thomson[1] for radiotellurium and by Rutherford[2] for radium.

[1] J. J. Thomson: Proc. Camb. Phil. Soc., Nov. 14, 1904.

[2] Rutherford: Phil. Mag., Aug., 1905.

Miss Slater[1] has recently shown that the emission of α particles from the thorium and radium emanations is also accompanied by slow moving electrons carrying a negative charge. The expulsion of such electrons is probably not a true radiation from the active matter itself, but is largely a secondary effect produced when the α particles impinge upon or escape from matter. It is for this reason not advisable to call them β rays, for this name should be retained for the primary β particles emitted from radioactive substances with velocities approaching that of light. J. J. Thomson has suggested that the name δ rays be applied to such slowly expelled electrons.

In the next chapter, we shall show that the changes in radium do not end with radium C, but continue through three more distinct stages. The calculations adopted to analyze the active deposit of rapid transformation are not, however, appreciably affected by the presence of these further products, for the activity due to them is in most cases less than one millionth of that observed on the active body immediately after removal from the emanation.

[1] Miss Slater: Phil. Mag., Oct., 1905.

CHAPTER V

ACTIVE DEPOSIT OF RADIUM OF SLOW TRANSFORMATION

A BODY which has been exposed in the presence of the radium emanation and then removed does not completely lose its activity. A small residual activity is always observed, the amount depending not only on the quantity of the emanation to which it has been exposed, but also upon the time of exposure. This small residual activity was first observed by Mme. Curie and has been closely examined by the writer.

After removal from the emanation, the activity of a body at first decays according to the laws discussed in the last chapter. There is finally an exponential rate of decay with a period of 26 minutes. Twenty-four hours after removal, the active deposit of rapid transformation has disappeared almost completely, and the activity left behind is generally less than one millionth of the activity observed immediately after removal from the emanation.

In the present chapter, the variations of this activity with time will be considered and the changes occurring in the matter deduced. The active deposit of slow transformation will be shown to consist of three successive products called radium D, E, and F. The analysis of this apparently insignificant residual activity observed on bodies has yielded results of considerable importance. It has disclosed the origin of the radiolead of Hofmann, of the radiotellurium of Marckwald, and also of the polonium of Mme. Curie; for these substances will be shown to be derived from the transformation of the radium atom.

It might at first sight be thought that this slight residual activity observed in bodies was due, not to the deposit of an active substance upon them, but to a possible effect produced by the powerful radiations of the emanation to which the bodies had been exposed.

This point was examined by the writer[1] in the following way:

The interior surface of a glass tube was covered with equal areas of thin metal, including platinum, aluminium, iron, copper, silver, and lead. A large quantity of emanation was introduced into the tube and left there for seven days. The activities of the plates, two days after removal from the emanation, were separately tested, and found to be unequal, being greatest for copper and silver and least for aluminium.

After standing for another week, the initial variations of activity had largely disappeared. These initial differences of activity were due to slight differences in the rates of absorption of the radium emanation by the metals. As this emanation was released, the activities of the plates reached equal values. The radiations from each plate consisted of α and β rays and were identical in penetrating power. This result shows that the residual activity observed cannot be due to any direct actions of the radiations on the body, for if this were the case, we should expect the activity of the different metals to vary not only in quantity but in quality. We may conclude then, that the activity is due to an active substance deposited on the surface of the metals. This view is completely borne out by later experiments, for it will be shown that the active deposit can be removed from a platinum plate by solution in acids and can also be volatilized at a high temperature.

Variation of the α Ray Activity with Time

The α ray activity of the body after reaching a minimum value during the first few days, steadily increases in amount for several years. The activity during the first few months increases nearly proportionally with the time. The curve (Fig. 33) then begins to bend over, and after 240 days — the period over which it has so far been examined — becomes much

[1] Rutherford: Phil. Mag., Nov., 1904.

more flattened and obviously approaches a maximum value. The explanation of this rise of activity will be considered at a later stage.

Variation of the β Ray Activity with Time

The residual activity initially comprises both a and β rays, the latter being present in a much greater relative proportion

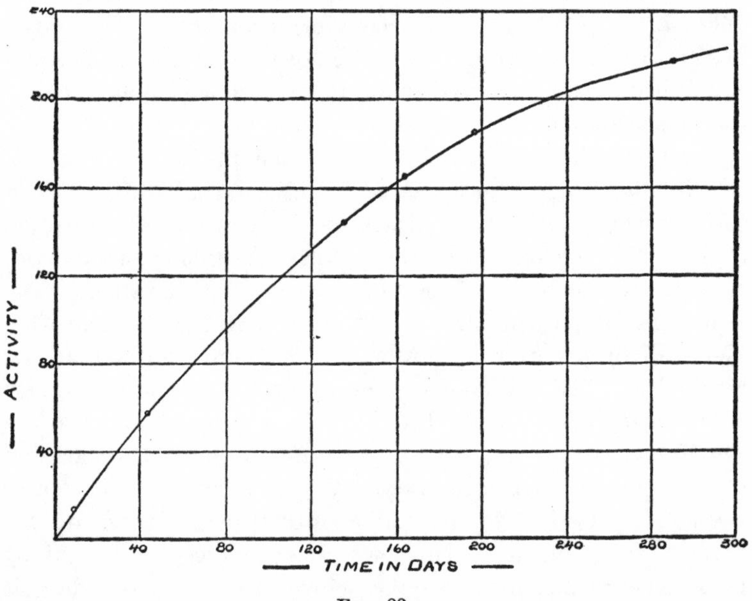

Fig. 33.

Rise of a ray activity of a body after exposure to the radium emanation. The activity is a measure of the amount of radium F present.

than is observed in radium or uranium. The β ray activity is small at first, but increases with time, reaching a maximum after about 50 days. The variation of activity with time is shown in Fig. 34. A plate was exposed for 3.75 days to the radium emanation, and the observations of the β ray activity, by means of an electroscope, were begun 24 hours after removal. The time was measured from the middle of the period of exposure to the emanation. The curve is seen to be similar in shape to the recovery curve of the emanation or of ThX. The β ray activity I_t at any time t after removal is given by

$$\frac{I_t}{I_0} = 1 - e^{-\lambda t}.$$

The activity reaches half value in about 6 days, and after 50 days has nearly reached a maximum.

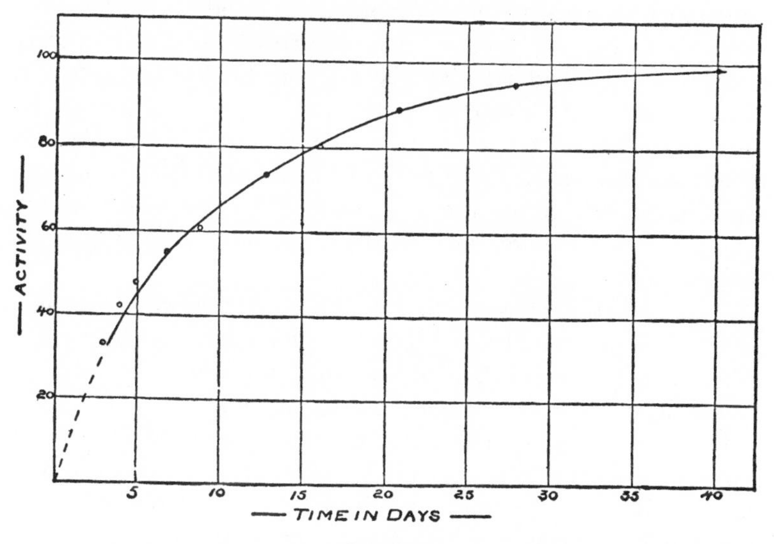

FIG. 34.

Rise of β ray activity of a body after exposure to the radium emanation. The β ray activity is a measure of the amount of radium E present.

Observations of the β ray activity were continued for 18 months, but showed that the activity remained practically constant after 50 days.

A curve of this character indicates that the β ray product is produced at a constant rate from a primary source, whose rate of transformation is so slow as to appear nearly constant over the period of observation. It follows from the curve that the β ray product has a period of 6 days.

The fact that the α and β ray activities increase almost from zero shows that their primary source, called radium D, is a rayless product, which, as we shall see later, is probably half transformed in about 40 years. Radium D is transformed into the β ray product called radium E, which is half transformed in about 6 days.

Effect of Temperature on the Activity

A platinum plate several months after its removal from the emanation was placed in an electric furnace and heated for a few minutes to varying temperatures. Exposure for four minutes, first at 430°C. and later at 800°C., had little if any effect in altering either the α or β ray activity. The α ray activity was, however, almost completely removed by heating the plate to about 1050°C., while the β ray activity did not at the time show any change. This result shows clearly that the product which emits α rays is more volatile than the product which emits β rays.

This experiment is another example of the way in which differences in volatility of two products may be utilized to effect a partial separation of one from the other.

We now come to another striking observation. The β ray activity of the platinum plate, though apparently unchanged immediately after the heating, began slowly to decrease, and finally fell to one quarter of the initial value. Subtracting this residual activity, it was found that the β ray product lost its activity exponentially, falling to half value in about 4.5 days.

We may thus conclude that the heating of the active deposit had a double action; for not only was the α ray product (which will be shown to arise from the β ray product, radium E) driven off, but about three quarters of the primary source, radium D, was also volatilized.

We thus have the striking result that in a mixture of three successive products, the first and third are mostly volatilized at a temperature of about 1000°C., while the middle product is unaffected. It will be observed that the period of decay of the β ray product (4.5 days), observed after heating, does not agree with the period of the same product (6 days) deduced from the rise curve of Fig. 33. This difference requires further investigation. The 6 day period is probably the more correct value under normal conditions.

Separation of the α Ray Product by Bismuth

The emanation from 30 milligrams of radium bromide was condensed in a glass tube and left there for one month. The active deposit remaining on the surface of the glass was then dissolved in dilute sulphuric acid and the solution laid by for about a year. During this interval the α ray activity steadily increased. By introducing a polished bismuth disc into the solution, the α ray product may be deposited electrochemically on the bismuth. By introducing several bismuth discs successively into the solution and allowing them to remain for several hours, the α ray product was mostly removed. On evaporating the solution to dryness, it was found that only 10 per cent of the original α ray activity remained behind.

The β ray activity of the solution was not altered by this process. The bismuth discs gave out only α rays, but no trace of β rays. This result shows that only the α ray product was removed. Radium D, as well as E, was left behind, for if some radium D had been deposited on the bismuth, it would have changed into radium E, and consequently some β rays would have been emitted from the bismuth disc, after standing for several weeks in order to allow time for D to change into E.

No such effect, however, was observed. The activities of these bismuth discs were tested in an α ray electroscope for over 200 days. The activity of each was found to fall off nearly exponentially, reaching half of the initial value after about 143 days. We may thus conclude that the substance which emits α rays is a simple product, which is half transformed in 143 days. This α ray product will be called radium F, for it will be shown to be a successive product of radium E.

The fact that radium E is the parent of F is clearly brought out by the following experiment.

A platinum wire coated with the active deposit of slow transformation was exposed for some minutes to a temperature of over 1000°C. Most of the radium F was volatilized. The α ray activity of this platinum plate was then carefully examined for several weeks. The small α ray activity, observed immediately after removal from the furnace, increased rapidly during the first two weeks and then more slowly. The gain of α ray activity with time is shown in Fig. 35.

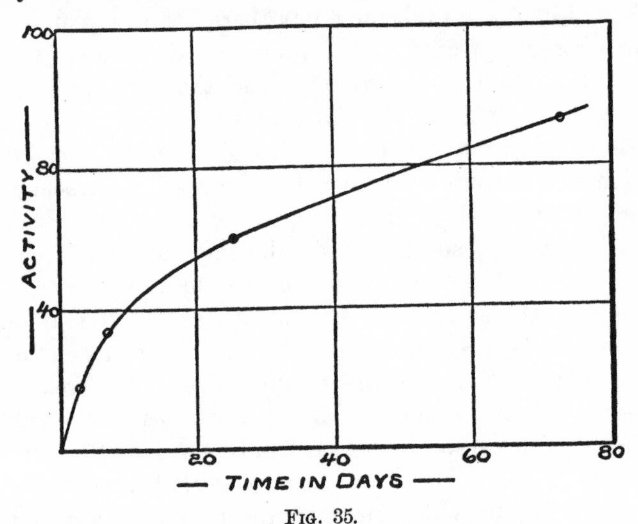

FIG. 35.

Rise of α ray activity on a platinum plate which has been heated to a temperature sufficient to remove most of the radium D and F present. Radium E is left behind and changes into F.

A curve of this character is to be expected if radium E is the parent of F. The action of the high temperature volatilized the greater part of D and F, but left E behind. The radium E was then transformed with a period of 4.5 days and changed into F. The α ray activity thus initially rose rapidly, due to the new radium F supplied by the transformation of E. The slow increase observed after some weeks, when most of the radium E present had been transformed, was due to the production of radium F by the small amount of D and E which was not volatilized from the platinum plate.

We may thus conclude that radium E is the parent of radium F.

It has been previously shown that radium E is produced from radium D, which itself does not emit β rays. The small α ray activity observed initially when radium D is present in maximum amount shows that radium D does not emit α rays. Radium D is consequently a rayless product.

Summary of the Radium Products

The analysis of the active deposit of slow change has thus disclosed the existence of three successive products of radium. The period of change of these products and some of their distinctive physical and chemical properties are tabulated below.

Product.	Time to be half transformed.	Rays.	Chemical and physical properties.
Radium D	About 40 years	None	Soluble in strong acids, volatilized at or below 1000° C.
" E	6 days	β and (γ?)	Non-volatile at 1000° C.
" F	143 "	α	Volatile at about 1000° C., deposited from a solution on a bismuth plate.

The method for deducing the period of radium D will be considered a little later. A sufficient amount of radium E has not yet been collected to test whether it gives out γ as well as β rays. But since in every other substance examined, these two types of rays always go together, it is almost certain that radium E gives out γ rays.

In the last chapter it was shown that the active deposit of rapid change consists of the three successive products, radium A, B, and C. It is thus natural to conclude that radium D is derived directly from the transformation of radium C. It is difficult to show definitely that radium C is the parent of D. We know, however, that D must be derived from either the emanation or one of its products, and since the products A, B, and C are lineal descendants of the emanation, the most plausible assumption is that the family of products D, E, and F are also lineal descendants of radium C.

On this assumption, the various radium products, together

with their periods of transformation and the types of rays emitted, are shown diagrammatically below (Fig. 36).

It is instructive for a moment to review briefly the series of changes exhibited by radium. The radium atom is a comparatively stable one, and on an average only half the radium atoms break up in 1300 years. The a particle projected during the disintegration of the atom has a velocity slower than those from the radium products, and can only pass through 3.5 cms. of air before complete absorption. The radium suffers a radical change on account of the loss of the a particle, and is transformed into a gas — the radium emanation — which is far more unstable than radium itself, for half of it breaks up in 3.8 days. After the expulsion of an a particle, the product radium A makes

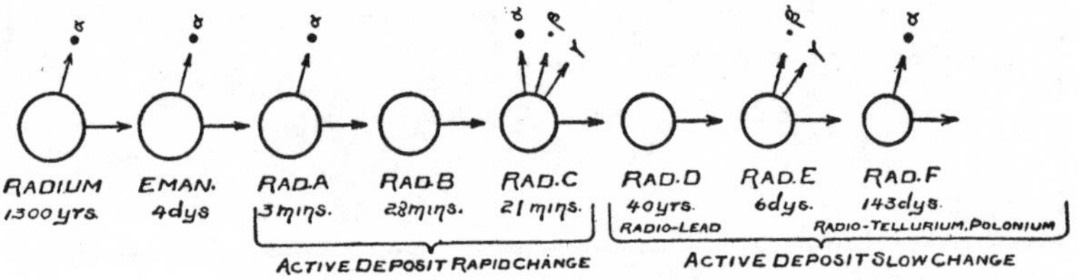

Fig. 36.

Radium and its family of products.

its appearance. This is the most unstable of all the radium products, for half of it breaks up in 3 minutes.

The next product is radium B, with a period of 26 minutes. It has the peculiarity of being transformed without the emission of rays at all. This points either to a transformation by the rearrangement of the components of the atom without any loss of mass, or, as is more likely, to the emission of an a particle at a velocity too low to ionize a gas. It will be seen later that the a particle loses the property of ionization when its velocity falls below about one fortieth of the velocity of light, so that the a particle may be expelled at a considerable speed and yet show no ionization effects. The next substance is radium C, which is the most remarkable of all the radium products, for in breaking up it emits all three types of rays. It would appear as if the transformation of C were accompanied by a most violent explosion in the atom, for not only is the α particle ejected with a greater speed than from any of the other products, but at the same time β particles are expelled with a velocity nearly equal to that of light. There is also an emission of very penetrating γ rays.

The α particle projected from radium C can traverse 7 cms. of air before complete absorption, while the α rays from the other products have a range not greater than 4.8 cms. After this violent atomic outburst, the residual atom, radium D, is far more stable and breaks up without the appearance of rays.

The next product is radium E, which emits only β and γ rays. It has a comparatively short life, but gives rise in turn to radium F, which has a slow period of change. No further products of transformation have been detected, and the interesting question of the final or end product of radium is reserved for discussion in Chapter VIII.

Period of Change of Radium D

Radium D does not emit rays, and consequently neither its properties nor its rate of transformation can be determined by direct means. The following product, radium E, however, emits β rays, and by noting the variations of its activity when in equilibrium we should be able to detect any variation in the rate of change of the parent product D.

This is readily seen to be the case, for when equilibrium is reached, the number of atoms of E which break up per second will always be equal to the number of atoms of D breaking up per second. Unfortunately the rate of transformation of D is so slow that no certain change in the equilibrium activity of E has been detected in the course of one year, and a long interval of time will probably be necessary to fix the period of D by direct measurement.

It is of importance to form a rough estimate of its probable period. This can be deduced on certain assumptions which are probably approximately realized in practice.

Suppose that a quantity of emanation is introduced into a

closed vessel and left there to decay. Several hours after the introduction of the emanation, the amount of radium C which emits β rays reaches a maximum value, and then decays at the same rate as the emanation. If q_1 is the maximum number of β particles emitted per second from radium C at its maximum activity, the total number N_1 emitted during the life of the emanation is very approximately given by $N_1 = \frac{q_1}{\lambda_1}$, where λ_1 is the constant of change of the emanation. Suppose that the active deposit of slow transformation is allowed to remain undisturbed for about 50 days after the emanation has practically disappeared. Radium D and E will then be in equilibrium. Let q_2 be the number of β particles emitted from D and E. Then if the transformation of radium D follows the ordinary exponential law with a constant λ_2, the total number N_2 of β particles emitted during the life of radium D is given as before by $N_2 = \frac{q_2}{\lambda_2}$. But if each atom of radium C in breaking up emits one β particle, the total number of β particles emitted during the life of the emanation must be equal to the number of atoms of emanation originally present. The number of atoms of D formed by the emanation will also be equal to this quantity, and if each atom of D gives rise to one of E, which breaks up with the expulsion of one β particle, we see that the total number of β particles expelled from C during the life of the emanation must be equal to the total number of particles expelled from E during the life of D. Consequently $N_1 = N_2$, and therefore

$$\frac{\lambda_2}{\lambda_1} = \frac{q_2}{q_1}.$$

It is not easy to measure directly the number of β particles expelled either from radium C or E, but on the assumption that the average β particle emitted from C or E produces about the same ionization in a gas,

$$\frac{q_2}{q_1} = \frac{i_2}{i_1},$$

where i_1, i_2 are the saturation ionization currents due to C and E respectively, measured under the same conditions in the same testing vessel. This ratio, $\frac{i_2}{i_1}$, can be readily determined, so that the ratio $\frac{\lambda_2}{\lambda_1}$ is known. Substituting the value of λ_1 for the emanation, λ_2 can be determined.

Proceeding by this method, the writer[1] deduced that radium D should be half transformed in about 40 years. This period is almost certainly of the right order, but from the nature of the assumptions, the value cannot pretend to be more than a first approximation to the truth. The main source of error probably lies in the assumption that the β particles of radium C and E produce the same average ionization in the gas.

As a criterion of the order of accuracy obtained in predicting periods of change by these means, it may be mentioned that, by a similar method, I deduced that the period of radium F was about one year. Actual observation has since shown that this period is 143 days. I think that the period of D will certainly be found to lie between 20 and 80 years.

Variation of the α and β Ray Activity over Long Periods of Time

We are now in a position to deduce the variation of the α and β ray activity for the active deposit over long intervals of time. Since radium E is transformed at a rapid rate compared with F, we may assume as a first approximation that D is transformed directly into F. The problem thus reduces to the following: Given that the periods of two successive products are 40 years and 143 days respectively, find the number of atoms of each product present at any time. This is exactly equivalent to the practical case already considered on page 50 for the active deposit of thorium where the two changes had periods of 11 hours and 55 minutes respectively.

The β ray activity of D and E after reaching its maximum will decrease exponentially, falling to half value in 40 years.

[1] Rutherford, Phil. Mag., Nov., 1904.

Using the equation discussed on page 51, it can at once be deduced that the number of atoms of radium F reaches its maximum in about 2.6 years and that this substance will ultimately decay *pari passu* with the parent product D, *i. e.*, it will be half transformed about 40 years later. The curves shown in Fig. 37 give the relative number of atoms of E and F which break up per second at any time after the formation of the

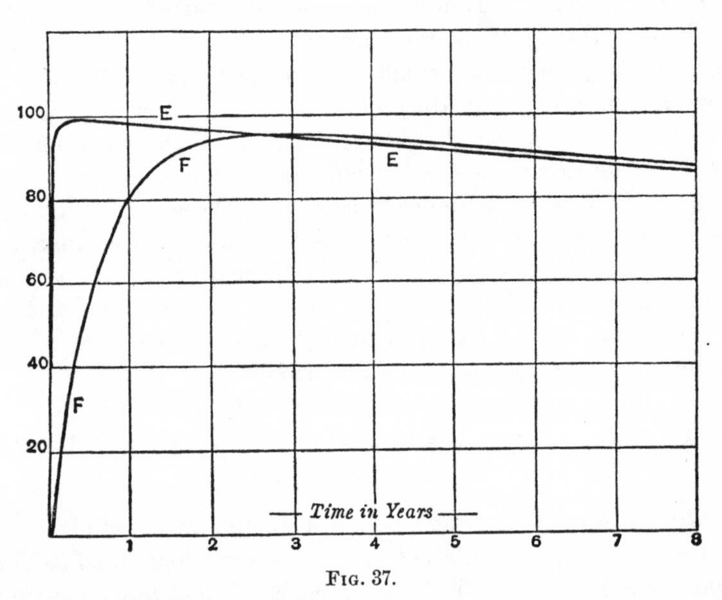

FIG. 37.

Curve E E represents the variation in the number of atoms of radium D breaking up per second. Curve F F represents the number of atoms of radium F breaking up per second.

deposit. Since the activity of F is proportional to the number of atoms of F which break up per second, we see that the activity of F will rise from zero to a maximum after 2.6 years, and will then decay with a 40 year period.

The variation of the a ray activity with time, is in good agreement with the theoretical curve over the range so far examined (See Fig. 33).

It is of interest to note that the same a ray activity observed

9 days after the formation of the active deposit is again reached after an interval of about 180 years.

The production by radium of the active deposit of slow transformation at once explains the strong radioactivity observed in rooms in which large quantities of radium have been used, even when the radium has been completely removed for some time. This effect has been observed by several experimenters, and especially by those who have been occupied in the separation and concentration of large quantities of radium.

The radium emanation released from the radium is transferred by diffusion and convection currents throughout the whole laboratory, and far distant rooms into which radium preparations have not been introduced become permanently radioactive. The emanation is transformed *in situ* through the succession of products radium A, B, and C, and finally passes into the active deposits of slow change. This matter is deposited on the interior surface of rooms and on every object in the building. For a given supply of emanation, the α ray activity will be small at first, but will steadily increase for about three years.

This residual activity on bodies is a source of considerable disturbance in radioactive work. Eve,[1] for example, found that every substance examined in the Macdonald Physical Laboratory of McGill University showed an abnormally large natural activity. At the time of testing, this activity was about 60 times greater than that observed in the same laboratory before the introduction of large quantities of radium into the building. All electroscopes made of materials exposed in the building had a large natural leak due to the active deposit. This can in part be removed by cleaning with sandpaper or by solution in acids. Unless all electroscopes or testing vessels are made outside the laboratory to insure a small natural leak, measurement of very weak radioactivities is rendered almost impossible. When once a building has been infected, it does not serve any immediate purpose to remove the radium, for the

[1] Eve: Nature, March 16, 1905.

a ray activity will continue to increase for about three years and will last for hundreds of years.

For these reasons it is very advisable to reduce the escape of emanation into the air of a laboratory as far as possible and to keep all radium salts in sealed vessels.

Presence of the Active Deposit in Radium

Since radium D is produced from radium at a nearly constant rate, it should gradually increase in quantity with the age of the radium. The presence of radium D in old radium can be detected in a very simple way. With a freshly prepared sample of radium, continued boiling for five or six hours removes the emanation as fast as it is formed, and reduces the β ray activity arising from radium C to a fraction of one per cent of its maximum value when in radioactive equilibrium.

A very different result is observed if an old preparation of radium is treated in a similar way. The writer had in his possession a small quantity of impure radium kindly presented by Professors Elster and Geitel four years before.

After continued boiling, the activity could not be reduced below 8 per cent of the original amount, or about 9 per cent of the activity due to C alone. This residual β ray activity was due to the radium E stored up in the compound.

The amount of radium E will steadily increase with time, and will reach a maximum value when the same number of atoms of radium C and radium E break up per second. The number of β particles expelled per second from radium E will, under such conditions, be equal to the number expelled from radium C. Since the radium itself is transformed very slowly compared with radium D, the amount of radium D produced per year (measured by the β ray activity of radium E) should be about 1.7 per cent of the equilibrium amount.

The β ray activity due to radium E should thus be about 7 per cent of that due to radium C after the lapse of four years. The observed and calculated values (9 and 7 per cent respectively) are thus in fair agreement.

By adding a trace of sulphuric acid to the radium solution,

the radium was precipitated and the products D, E, F, which are soluble in sulphuric acid, remained in the solution. The filtrates thus contained a large part of the above three products. The radium F was removed from the solution by means of bismuth discs, and showed an activity to be expected from the age of the radium.

Variation of the Activity of Radium with Time

We shall see later that radium itself, apart from its products, is probably half transformed in about 1300 years, and consequently the number of atoms breaking up per second decreases exponentially according to this period. In consequence of the formation of the active deposit of slow change, the activity supplied by it will at first more than compensate for the decrease in the activity of the radium itself. The activity will rise for several hundred years, but will ultimately decay exponentially with the period of the radium.

When sufficient time has elapsed for approximate equilibrium between the mixture of products, the number of β particles expelled from the old radium will be twice that due to radium C alone; for radium E will emit per second, under such conditions, the same number of β particles as radium C.

It can readily be calculated from the theory of two changes, that the number of β particles expelled from radium and its products will steadily increase until a maximum is reached after 226 years. After that period, the number will decrease nearly exponentially with a period of 1300 years.

The variation with time of the number of β particles expelled from radium is shown in Fig. 38, Curve BB.

Radium and its family of rapidly changing products together emit four α particles for the one emitted from radium F. By calculation it can be shown that the number of α particles expelled from radium will reach a maximum after about 111 years, and will then be about 1.19 times the number emitted from radium about one month old. The number, as in the case of the β particles, will then decrease and the period will be 1300 years.

Curve AA shows the variation with time of the number of *a* particles expelled from radium and its mixture of products. Curve CC represents the number of radium atoms breaking up per second.

These calculations of the variation of the activity of radium with time depend upon the accuracy of the periods of change of

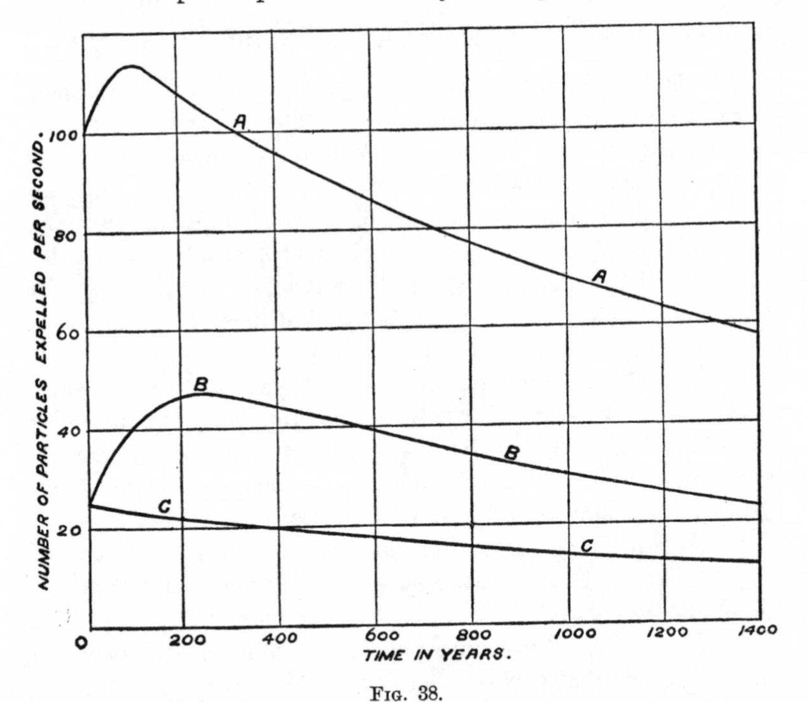

Fig. 38.

Curve A A represents the variation with time in the number of *a* particles expelled from radium. Curve B B represents the number of *β* particles expelled per second. Curve C C represents the number of atoms of radium breaking up per second.

radium and radium D. Any alteration in these values will to some extent alter the curves of variation of activity with time.

Identity of Radium F with Radiotellurium

Since the products D, E, and F are continuously produced by radium, they should be found in all radioactive minerals containing radium, and in amounts proportional to the amount of radium in the mineral. It is now necessary to consider whether any of these products have been previously separated from radioactive minerals and known by other names.

We shall first consider the product radium F, which will be shown to be identical with the very active substance called radiotellurium, separated by Marckwald from pitchblende residues. In endeavoring to establish the indentity of two products, the main criteria to be relied upon are: —

(1) the identity of the radiations or characteristic emanations emitted by the products;

(2) the identity of the periods of change of the products;

(3) the similarity of chemical and physical properties of the active products.

The third criterion is initially of less importance than (1) or (2), since in most cases the active products are separated in a very impure state and the apparent chemical reaction may be largely modified by the presence of impurities.

We have seen that the product radium F emits only α rays, has a period of about 143 days, and is deposited on bismuth from the active solution. Radiotellurium behaves in identically the same manner. In addition, the writer[1] directly compared the rates of decay of the activities of radiotellurium and radium F and found them to be the same within the limit of experimental error. Each loses half its activity in about 143 days. The period of decay of radiotellurium has also been experimentally examined by Meyer and Schweidler and Marckwald. The former found a period of 135 days and the latter 139. Considering the difficulty of making accurate comparative measurements over such long intervals of time, the values obtained by the different observers are in remarkably good agreement.

The writer also found that the rays emitted from radium F had the same penetrating power as those emitted from an active bismuth plate coated with radiotellurium. It is known from the work of Bragg and others, that each product of radium emits α rays of a penetrating power, which is definite for each,

[1] Rutherford: Phil. Mag., Sept., 1905.

but varies considerably among the different products. This equality in penetrating power thus supplies strong evidence in favor of the identity of the two products.

We may thus conclude that the radiotellurium of Marckwald contains as its active constituent the product radium F; or, in other words, radiotellurium is a transformation product of radium.

The methods of separation and concentration of radiotellurium used by Marckwald are of special interest. The separation by Mme. Curie of radium from pitchblende in which it existed in the proportion of less than one part in a million was in itself a notable performance, but the work of Marckwald in the separation of radiotellurium constitutes a still more striking illustration of the possibility of chemically concentrating a radioactive substance existing in almost infinitesimal amount.

Marckwald initially observed that a bismuth rod dipped into a solution of pitchblende residues became coated with a deposit which emitted only *a* rays. After some days, the active substance was in this way almost completely removed from the solution and obtained on the bismuth. The deposit on the bismuth was found to consist for the most part of tellurium, and for this reason Marckwald called the active substance radiotellurium. Later Marckwald devised very simple and efficient means of separating the active matter from the tellurium, and finally obtained a substance which, weight for weight, was far more active than radium.

Five tons of uranium residues, corresponding to 15 tons of the Joachimsthal mineral, were worked up to extract the radiotellurium from it, and he finally obtained only 3 milligrams of the active substance. If plates of tin, copper, or bismuth were dipped into a hydrochloric acid solution of this substance, they were found to be covered with a finely divided deposit. These plates were extremely radioactive, and gave marked ionizing, photographic, and phosphorescent effects. As an illustration of the enormous activity of this substance, Marckwald states that a weight of 1/100 of a milligram deposited on a copper plate 4 sq. cms. in area lighted up a zinc sulphide screen brought near it so strongly that the luminosity could be clearly seen by an audience of several hundred people.

In consequence of the minute amount of working material, Marckwald has not yet succeeded in purifying the active substance sufficiently to determine its spectrum.

By means of a simple calculation, the activity of radium F, *i.e.* of radiotellurium in a pure state, can readily be deduced. Let N_1 be the number of atoms of radium F in one gram of the radioactive mineral, and N_2 the number of radium atoms. Radium and radium F are in radioactive equilibrium, and consequently the same number of atoms of each break up per second.

Thus, $$\lambda_1 N_1 = \lambda_2 N_2,$$

where λ_1, λ_2 are the constants of change of radium F and radium respectively. Now radium F is half transformed in .38 years and radium in about 1300 years. Consequently

$$\frac{N_1}{N_2} = \frac{.38}{1300} = .00029.$$

Now it is probable that the atomic weights of radium and radium F are not very different. Consequently, for every gram of radium in the mineral, there exists only .29 milligram of radium F. For equal weights, the number of a particles expelled from radium F is 3400 times as great as the number expelled from radium itself, or 850 times as great as the number expelled from radium about one month old, when it is in equilibrium with its three rapidly changing a ray products.

Assuming that the a particle from radium F produces about the same amount of ionization as the average a particle from radium, the activity of radium F measured by the electric method should be 850 times greater than that of radium.

It has been found experimentally that the amount of radium in radioactive minerals is always proportional to their content of uranium and that for every gram of uranium there is present 3.8×10^{-7} gram of radium.

The weight of radium F per gram of uranium is thus 1.1×10^{-10} gram, and per ton of 2000 lbs., 0.1 milligram. From

15 tons of Joachimsthal mineral, which contains about 50 per cent of uranium, the yield of radium F should be 0.75 milligrams.

The amount separated by Marckwald from this amount of pitchblende was about 3 milligrams. It is unlikely that the whole amount of radium F was separated, and the three milligrams probably contain some impurity. The theoretical proportion of radium F in the radioactive mineral is thus in good agreement with the experimental results.

Although the proportion of radium F in minerals may appear extremely small, yet the other more rapidly changing products exist in still smaller amounts. The weight of each product present per ton of uranium is directly proportional to its period, so that the most swiftly changing product is present in the smallest quantity. In the following table are shown the weights of each of the radium products present per ton of 2000 pounds of uranium in the mineral.

Products.	Period.	Weight in milligrams per ton of uranium.
Radium	1300 years	340 milligrams
Emanation	3.8 days	2.6×10^{-3} milligrams
Radium A	3 minutes	1.4×10^{-6} "
" B	26 "	1.2×10^{-5} "
" C	19 "	9×10^{-6} "
" D	40 years	10 "
" E	6 days	4.2×10^{-3} "
" F	143 "	.1 "

The products radium A, B, C, and E exist in far too small amounts to be examined by ordinary chemical methods, even if their short life allowed it. Radium D, however, is present in considerable quantity compared with radium F, and it should be possible to obtain a sufficient amount of it for a chemical examination.

Polonium and Radiotellurium

It will be remembered that the first active substance separated from pitchblende was found associated with bismuth and was called polonium by its discoverer, Mme. Curie.

Several methods were devised for the concentration of the active material mixed with the bismuth, and Mme. Curie finally succeeded in obtaining an active substance comparable in activity with radium. The polonium gave out only α rays, and its activity was not permanent but gradually decreased.

Both as regards the nature of its rays and its physical and chemical properties, polonium is very analogous to the product radium F and radiotellurium. There has been a considerable amount of discussion at various times as to whether the active constituent in radiotellurium is identical with that present in polonium. At first it was announced that the activity of radiotellurium did not decay appreciably, and it apparently behaved in this respect quite differently from polonium. We now know that radiotellurium does lose its activity, and fairly rapidly.

If the two products contain the same constituent, their activities should decay according to the same period. Mme. Curie, however, has observed that some of her preparations of polonium do not lose their activity according to an exponential law.

For example, a sample of polonium nitrate lost half its activity in 11 months and 95 per cent in 33 months. A sample of the metal lost 67 per cent of its activity in 6 months. These results are not at all concordant. The sample of the metal loses its activity slightly faster than radium F, while the nitrate at first loses it much more slowly. If these results are reliable, the divergence of the activity curves from the exponential law shows that more than one substance is present in the polonium experimented with by Mme. Curie. It is very probable that this second constituent is radium D. The presence of some of this substance, which gives rise to radium F, would cause the α ray activity to decrease at first more slowly than the normal rate when only radium F is present.

Considering the similarity of polonium and radiotellurium in their chemical, physical, and radioactive properties and the probable identity of their periods, I think that there can be no doubt that the α ray constituent in polonium is the same as that separated by a different method by Marckwald. We may then conclude that radiotellurium and polonium are both derived from the transformation of the radium atom.[1]

Connection of Radiolead with the Active Deposit

We shall now describe some experiments which show conclusively that the product radium D is the primary constituent of the radiolead, first separated by Hofmann from pitchblende residues. The early results of Hofmann, on the separation and properties of radiolead, were subjected to considerable criticism, but there is now no doubt that to him belongs the credit of separation of a new product from pitchblende, which proves to be the parent of radiotellurium and polonium.

My attention was first drawn to the connection between radiolead and the active deposit of radium by an examination of a specimen of radiolead kindly prepared for me by Dr. Boltwood of New Haven. This was found to give out initially an unusual proportion of β rays as compared with its α ray activity, and the α ray activity was found to increase progressively with time. In these respects it behaved in a similar manner to a substance initially containing radium D and E, in which the product radium F was being gradually formed, thus giving rise to the increasing α ray activity.

The connection of radiolead with radium D, E, and F has been conclusively proved by a chemical examination of the radioactive constituents found in radiolead and a determination of their periods of decay. It must be borne in mind that the name radiolead was given to the active substance because it was first separated mixed with lead, but we now know that the active substances contained in it have no more connection with lead than radium has with the barium from which it is finally separated.

[1] The identity of the active constituent of radiotellurium and polonium has now been definitely settled. Mme. Curie (Comptes rendus, Jan. 29, 1906) has accurately determined the loss of activity of polonium, and found that it decayed according to an exponential law with a period of 140 days. This period of decay is practically identical with that found for radiotellurium and radium F.

Hofmann, Gonders, and Wölfl [1] in the course of a chemical examination of a specimen of radiolead obtained the following results. Experiments were first made on the effect of adding substances to a solution of radiolead and then removing them by precipitation. Small quantities of the platinum metals in the form of chlorides were left in the solution for several weeks and then precipitated by formalin or hydroxylamine. All of these substances after separation were found to give out α and β rays.

Most of the β ray activity disappeared in about six weeks, and the α ray activity in about one year. We shall see that the β ray activity is due to the separation of radium E, which decays to half value in 6 days, while the α ray activity is due to radium F. This conclusion is further confirmed by experiments on the effect of heat on the activity of these substances. At a full red heat, the α ray activity was lost in a few seconds. This is in agreement with experiments on radium F which is volatilized at about 1000° C.

Salts of gold, silver, and mercury, added to the solution of radiolead, were found to show only α ray activity. This is explained if radium F is alone removed. Bismuth salts, on the other hand, showed α and β activity, but the latter died away rapidly. This shows that bismuth removes both radium E and F.

The α ray activity of the radiolead is much reduced by precipitation of bismuth in the solution, but gradually increases again with time. This result is exactly what is to be expected if radiolead contains radium D, E, and F. Radium E and F are removed with the bismuth, while D is left behind, and in consequence there is a fresh growth of radium E and F.

The radioactive measurements made by Hofmann, Gonders, and Wölfl were unfortunately not very precise, but this want has been supplied by some recent careful measurements by Meyer and Schweidler.[2] If radiolead contains radium D, E, and F, the β ray activity due to E should decay to half value

[1] Hofmann, Gonders, and Wölfl: Ann. d. Phys , v, p. 615 (1904).
[2] Meyer and Schweidler: Wien Ber., July, 1905.

in 6 days, and the α ray activity due to F to half value in about 140 days.

These results have been completely confirmed by Meyer and Schweidler, who have accurately measured the rates of decay of the various products from radiolead. A series of palladium plates were immersed in the radiolead solution. After removal the activity consisted of α and β rays. The β ray activity decreased exponentially with the time, falling to half value in 6.2 days. The β ray product is thus identical with radium E. The α ray activity, after some months, fell off exponentially with a period of 135 days. The α ray product is thus identical with radium F.

There is thus no doubt that radiolead some time after its preparation contains radium D, E, and F. No observations have so far been made to settle definitely whether radium D, E, and F are removed together with the lead or whether only radium D is removed and the presence of radium E and F after some time is due to their production from radium D. If the bismuth is separated from the lead, it seems likely that radium E and F would be removed with the former and that radium D alone would be removed with the lead.

It is thus seen that the primary constituent in radiolead is the parent of radiotellurium and polonium.

The connection of the radium products with radiolead is outlined in the following table.

Radium D = product in new radiolead. No rays. Half transformed in 40 years.

Radium E gives out β rays; is separated with bismuth, iridium, and palladium. Half transformed in 6 days.

Radium F = product in polonium and radiotellurium. Gives out only α rays. Volatile at 1000° C., and attaches itself to bismuth and palladium. Half transformed in 143 days.

These results have thus emphasized the importance of radium D as a new radioactive substance in pitchblende.

It has been shown that about 10 milligrams of radium D should be separated from the mineral for each ton of uranium present. A few weeks after separation the β ray activity of this substance should be about 30 times as great as that of radium. A quantity of this substance would serve as a useful source of β rays, and also as a very convenient means of obtaining radium F. A very active deposit of this substance could at any time be obtained by placing a bismuth or palladium plate in the solution. It is to be hoped that this substance will be separated from pitchblende residues at the same time as radium, for in many respects it would prove as useful in experiments as radium itself.

CHAPTER VI

ORIGIN AND LIFE OF RADIUM

Since radium itself continuously throws off α particles and gives rise to a radioactive gas, its amount must steadily decrease with time. Radium, in this respect, must be considered as a radioactive product like the emanation, the only difference being its comparatively slow rate of change. A given amount of radium left to itself must ultimately disappear as such, and after a series of transformations there will only remain the inactive substances produced by its decomposition.

The time of observation has been far too short to fix the period of change of radium by direct experiment. Accurate measurements of the activity will not supply any information of value on this point for a long interval, since the slow transformation products arising from the radium actually cause a steady increase of activity for several hundred years.

There are several indirect methods which can be employed to deduce the probable period of radium, depending on (1) the number of α particles expelled per second, (2) the observed heating effect of radium, and (3) the observed volume of the emanation released from it.

Method 1. We shall first consider the method based on the rate of expulsion of α particles. By measuring the charge carried by the α rays expelled from a thin film of radium, the writer[1] found that the total number of α particles expelled per second from one gram of radium at its minimum activity was 6.2×10^{10}, assuming that each α particle carries with it the usual ionic charge of 3.4×10^{-10} electrostatic units. When in radioactive equilibrium with its family of rapidly changing

[1] Rutherford: Phil. Mag., Aug., 1905.

products, the number of expelled particles is four times as great.

The simplest assumption is that one α particle is expelled from each atom as it breaks up. Consequently 6.2×10^{10} atoms of radium break up per second. Now it has been found from data based upon experiment that one cubic centimetre of a gas, — hydrogen, for example, — at standard pressure and temperature contains 3.6×10^{19} molecules. From this it follows that one gram of radium of atomic weight 225 contains 3.6×10^{21} atoms of radium. The fraction of radium which breaks up per second is

$$\frac{6.2 \times 10^{10}}{3.6 \times 10^{21}} = 1.72 \times 10^{-11},$$

or 5.4×10^{-4} per year.

Like any other active product, the amount of radium must decrease according to an exponential law, so that the value of its constant of change λ is 5.4×10^{-4} (year)$^{-1}$.

From this it follows that half of the radium is transformed in about 1300 years. The average life of the radium atom which is measured by $1/\lambda$ is about 1800 years.

Method 2. The calculation of the life of radium can also be based on the observed heating effect of radium, which will be shown later (Chapter X) to be a direct measure of the kinetic energy of the expelled α particles. From measurements of the velocity and mass of the α particle expelled from radium, the average energy of motion, $\frac{1}{2}$ mv^2, of the α particle was found by the writer to be 5.9×10^{-6} ergs. Now it is found experimentally that one gram of radium emits heat at the rate of about 100 gram calories per hour. If this is due to the kinetic energy of the α particles, the number of such particles that must be expelled per second is about 2.0×10^{11}. The number from radium itself is one quarter of this. Using the same method of calculation as before, it is seen that half of the radium is transformed in about 1600 years — a value not very different from that deduced by the first method.

Method 3. We shall now consider the calculation of the

life of radium based on the observed volume of the emanation released from one gram of radium. Ramsay and Soddy found that this maximum volume was slightly greater than one cubic millimetre at standard pressure and temperature. Now one cubic millimetre of gas contains 3.6×10^{19} molecules. The number of molecules of emanation produced per second is λ times the equilibrium number present, where λ is the constant of change of the emanation. Assuming, as is probably the case, that the emanation is a monatomic gas, and that each atom of radium in breaking up gives rise to one atom of emanation, the number of atoms of radium breaking up per second is 7.6×10^{10}. Proceeding as before, this gives 1050 years as the period of radium.

The first two methods involve the assumption of the number of atoms present in one cubic centimetre of a gas. The calculation based on the volume of the emanation can, however, be made in a different way without this assumption. If one atom of radium by the loss of one α particle is changed into one atom of emanation, the molecular weight of the latter must be at least 200. The value deduced from experiments on diffusion is about 100, but on page 85 some reasons have been given in support of the view that this is an underestimate. One cubic millimetre of the emanation thus weighs as much as 100 c.mms. of hydrogen, *i. e.* 8.96×10^{-6} grams. The weight of emanation produced per second is λ times this amount, *i. e.* 1.9×10^{-11} grams. The weight of emanation produced per year is thus 6×10^{-4} grams, and this must be nearly equal to the weight of radium breaking up per year. This makes the period of radium about 1300 years.

Considering the uncertainty attaching to the exact values of the data used in these calculations, the periods deduced by the three methods are in good agreement. In calculations we shall take 1300 years as the most probable value of the period of radium.

Radium thus breaks up at a fairly rapid rate, and in the course of a few thousand years a mass of radium left by itself would lose a large proportion of its activity. Assuming that radium breaks up with a period of 1300 years, it can readily be calculated that after an interval of 26,000 years, only one millionth of the mass of radium would remain unchanged. If we suppose, for illustration, that the earth was originally composed of pure radium, the activity observed in the earth 26,000 years later would be about the same as that observed to-day in a good specimen of pitchblende. This period of years is very small compared with the age of the minerals of the earth, and unless the very improbable assumption is made, that the radium was in some way suddenly formed at a very late period in the earth's history, we are forced to the conclusion that radium must be continuously produced in the earth. It was early suggested by Rutherford and Soddy that radium might be a disintegration product of one of the radioactive elements present in pitchblende. Both uranium and thorium fulfil the conditions required as a possible parent of radium. Both have atomic weights greater than that of radium, and both are transformed at a very slow rate compared with radium. A cursory examination shows that uranium is the most likely parent, since radium is always found in largest amount in uranium minerals, while some thorium minerals contain very little radium.

We shall now consider some of the consequences that should follow if uranium is considered to be the parent of radium.

Several thousand years after the uranium has been formed, the amount of radium should reach a definite maximum value. Its rate of production by the uranium is then balanced by its own rate of disappearance. Under such conditions, the number of atoms of radium which break up per second is equal to the number of atoms of uranium which break up per second. Now, as far as observation has gone, uranium only emits one α particle during its transformation into uranium X. The product uranium X does not emit α rays, but only β and γ rays. On the other hand, we have seen that radium itself and four of its products, viz., the emanation, radium A, C, and F, emit α rays. The number of α particles expelled from the radium, and these products of its transformation, should thus be five times the number expelled from uranium. Assuming that the a particles from the radium products produce about the same ionization as the a particle from uranium, the activity of a radioactive mineral, which consists mostly of uranium, should be about six times that of uranium itself. Now the best pitchblende shows an activity about five times that of uranium, so that the theoretical result is approximately realized in practice. Until, however, the relative ionizations produced by the a particles from uranium and each of the radium products are accurately known, the relative activities to be expected cannot be fixed with certainty.

Another consequence of the theory is that the amount of radium in any radioactive mineral should be always proportional to its content of uranium. This must hold in every case, provided neither the uranium nor the radium have been removed from the mineral by physical or chemical action. This interesting question has been experimentally attacked by Boltwood,[1] Strutt,[2] and McCoy,[3] and has yielded results of the highest importance.

McCoy accurately compared the activities of different radioactive minerals, and showed that in every case the activity was very nearly proportional to their percentage content of uranium. Since, however, the radioactive minerals contain some actinium, and occasionally some thorium, these results indicate that the activity of all these substances, taken together, is proportional to the amount of uranium. Boltwood and Strutt employed a more direct method, by determining the relative content of uranium and radium in radioactive minerals. The amount of uranium was determined by direct chemical analysis, while the amount of radium was determined by measurements of the amount of radium emanation released by solution of the mineral. The relative amount of the latter can be determined with great accuracy by the electric method, which is the most convenient method of comparing quantitatively the amounts of radium in different minerals.

[1] Boltwood: Nature, May 25, 1904; Phil. Mag., April, 1905.
[2] Strutt: Trans. Roy. Soc. A., 1905.
[3] McCoy: Ber. d. d. chem. Ges., No. 11, p. 2641, 1904.

The results of both these observers show that there is a nearly constant ratio between the amounts of radium and uranium in every mineral examined, except in one case, which will be considered later. Minerals were obtained from various localities, both in Europe and America, which varied widely in chemical composition and in the percentage content of uranium. The experiments of Dr. Boltwood of Yale University, which have been made with great care and accuracy, show a surprisingly constant ratio between the amounts of uranium and radium.

A brief account will be given of the methods employed by him in his measurements. The percentage of uranium in the mineral under consideration was first determined by chemical analysis. A known weight of the finely powdered mineral was placed in a glass vessel A (Fig. 39), and sufficient acid introduced to dissolve it.

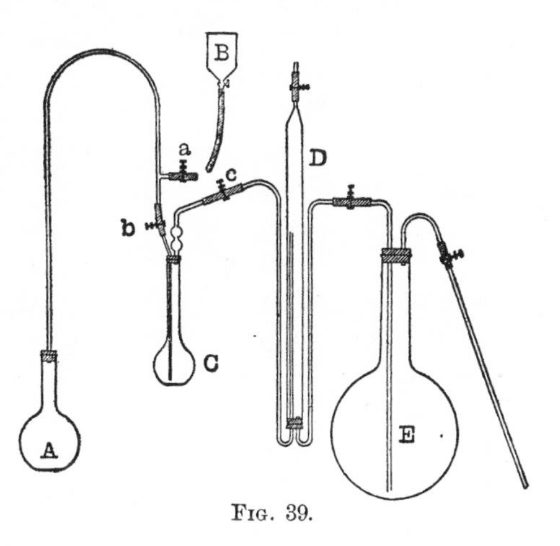

Fig. 39.

The acid was then boiled until the mineral was completely dissolved, and the emanation mixed with air was collected on the top of the column of water in the tube D. This emanation was then introduced into a closed electroscope of the type shown in Fig. 6, page 29, which was first exhausted. Air was then introduced until the gas inside the electroscope was at atmospheric pressure. On account of the excited activity produced by the emanation, the rate of discharge of the electroscope did not reach a maximum until about three hours after the introduction of the emanation. The rate of movement of the gold leaf of the electroscope was taken as a measure of the amount of emanation present. The emanations of thorium or actinium, released from the mineral at the same time as the radium emanation, had, on account of the rapid decay of their activity, completely disappeared before the introduction of the radium emanation into the electroscope. This process was repeated for all the minerals examined.

Boltwood observed that some minerals had considerable emanating power, *i. e.*, the minerals lost some of their emanation when in the solid state. Under these conditions, the amount of emanation released by solution and boiling of the mineral would be less than the equilibrium amount. The proper correction was made by sealing up a known weight of the mineral in a tube for one month and then measuring with the same electroscope the amount of emanation which collected in the air above the mineral. The sum of the two amounts gives the true equilibrium quantity of emanation corresponding to the radium present in the mineral.

The results obtained by Boltwood are shown in the following table. The numbers in column I give, in arbitrary units, the amount of emanation released by solution and boiling; column II shows the percentage of the emanation which escaped into the air; column III shows the amount of uranium in the mineral; and column IV the numbers obtained by dividing the equilibrium amount of emanation by the quantity of uranium present.

If the amount of radium always bears a definite ratio to the amount of uranium, the numbers in column IV should be the same. With the exception of some of the monazites, there is a remarkably good agreement, and, taking into consideration the great variation in the amount of uranium in the different minerals, and the wide range of locality from which they were obtained, the results afford a direct and satisfactory proof that the amount of radium in minerals is directly proportional to the amount of uranium present.

As an example of the confidence to be placed in this ratio as a physical constant for all radioactive minerals, Boltwood observed that some of the monazites contained a considerable quantity of radium, although the previous analyses had not shown any uranium to be present. A careful examination was made to

Substance.	Locality.	I.	II.	III.	IV.
Uraninite	North Carolina	170.0	11.3	0.7465	228
Uraninite	Colorado	155.1	5.2	0.6961	223
Gummite	North Carolina	147.0	13.7	0.6538	225
Uraninite	Joachimsthal	139.6	5.6	0.6174	226
Uranophane	North Carolina	117.7	8.2	0.5168	228
Uraninite	Saxony	115.6	2.7	0.5064	228
Uranophane	North Carolina	113.5	22.8	0.4984	228
Thorogummite	North Carolina	72.9	16.2	0.3317	220
Carnotite	Colorado	49.7	16.3	0.2261	220
Uranothorite	Norway	25.2	1.3	0.1138	221
Samarskite	North Carolina	23.4	0.7	0.1044	224
Orangite	Norway	23.1	1.1	0.1034	223
Euxinite	Norway	19.9	0.5	0.0871	228
Thorite	Norway	16.6	6.2	0.0754	220
Fergusonite	Norway	12.0	0.5	0.0557	215
Aeschynite	Norway	10.0	0.2	0.0452	221
Xenotine	Norway	1.54	26.0	0.0070	220
Monazite (sand)	North Carolina	0.88	. .	0.0043	205
Monazite (crys.)	Norway	0.84	1.2	0.0041	207
Monazite (sand)	Brazil	0.76	. .	0.0031	245
Monazite (massive)	Connecticut	0.63	. .	0.0030	210

test this point, and it was found that uranium was present in the amount to be expected according to theory. The failure to detect the presence of uranium in the earlier analysis was due to the presence of phosphates.

There is one interesting apparent exception to this constancy of the ratio between the amounts of uranium and radium. Danne recently found that considerable quantities of radium are present in certain deposits in the neighborhood of Issy l'Évêque in the Saône-et-Loire district, but that no trace of uranium could be detected. The active matter is found in pyromorphite (phosphate of lead), and in clays containing lead, but the radium is usually found in greater quantities in the former. The pyromorphite is found in veins of quartz and felspar rocks. The veins were always wet, owing to the presence of springs in the neighborhood. The percentage of uranium in the pyromorphite varies considerably for different specimens, but Danne states that on an average one centigram of radium is present per ton.

It seems probable that this radium has been deposited in the rocks after being carried from a distance by means of underground springs. The presence of radium in this district is not surprising, for crystals of autunite have been found about forty miles distant. This result is of interest, for it suggests that radium can in some cases be removed from the radioactive mineral by solution in water, and be deposited under suitable physical and chemical conditions some distance away. It also suggests the possibility that deposits containing a considerable proportion of radium may yet be discovered in positions where the conditions necessary for the solution and re-deposit of the radium are favorable.

Amount of Radium in Minerals

The weight of radium in a mineral, per gram of uranium, is thus a definite constant of considerable practical as well as theoretical importance. This constant was recently determined by Rutherford and Boltwood by comparison of the amount of emanation liberated from a known weight of uraninite with that released from a known quantity of pure radium bromide in solution. For the latter purpose, a known weight of radium bromide was taken from a sample of radium bromide obtained from the Quinin Fabrik, Braunschweig, which had previously been found to give out heat at a rate of over 100 gram calories per hour. P. Curie and Laborde found that their pure radium chloride preparations gave out heat at the rate of about 100 gram calories per hour. We may thus conclude that the radium preparation employed was nearly pure. This known weight — about one milligram — was dissolved in water, and by successive dilutions a standard solution was made up containing 10^{-6} grams of radium bromide per cubic centimetre. Taking the constitution of radium bromide as $RaBr_2$ and the atomic weight of radium as 225, it was deduced that in each gram of uranium in the mineral the corresponding weight of radium was 3.8×10^{-7} gram.[1]

From this it follows that .34 gram of radium is present in a mineral per ton of uranium. Since the radioactive minerals

[1] The first determination of this constant by Rutherford and Boltwood (Amer. Journn. Sci., July, 1905) gave a value of 7.4×10^{-7}. This was later found by them to be incorrect, owing to a precipitation of the radium in the standard solution.

from which radium is extracted usually contain about 50 per cent of uranium, the yield of radium per ton of mineral should be about .17 gram.

Assuming, as a first approximation, that the α particles from radium and its products, and from uranium, are expelled at the same speed, the activity of the radium and its family of rapidly changing products when in equilibrium with the uranium, should be four times that of uranium. Taking the activity of pure radium as about three million times that of uranium, the weight of radium required to produce this activity is

$$\frac{4}{3 \times 10^6} = 1.33 \times 10^{-6} \text{ grams.}$$

The observed amount, 3.8×10^{-7} grams, is considerably smaller. The agreement between theory and experiment, however, becomes much closer when we take into account the known fact that the average α particle from radium has a greater penetrating power and consequently produces a greater number of ions in the gas than the average α particle expelled from uranium.

Growth of Radium in Uranium Solutions

Although the constancy of the ratio between the amounts of radium and uranium in all radioactive minerals, as well as the agreement between the theoretical and observed quantities, afford very strong proof of the truth of the theory that uranium is the parent of radium, yet this conclusion cannot be considered as completely established until it has been experimentally shown that radium gradually collects in uranium solutions originally freed from it.

The rate of production of radium on the disintegration theory can readily be estimated. The fraction of radium breaking up per year has been calculated on page 149 and shown to be about 5.4×10^{-4} per year. The amount of radium per gram of uranium in minerals has been shown to be 3.8×10^{-7} grams. Consequently, in order to keep up the quantity of radium in a mineral at a constant amount, the rate of supply per year per gram of uranium must be $5.4 \times 10^{-4} \times 3.8 \times 10^{-7} = 2 \times 10^{-10}$

gram. This represents the amount of radium formed per year from each gram of uranium. The presence of radium can readily be detected by its emanation. Using a kilogram of uranium, the amount of radium formed per year is 2×10^{-7} gram. The emanation from this would cause a gold-leaf electroscope to be discharged in a few seconds, while the amount of radium produced in a single day should be easily measurable.

Experiments on the growth of radium in uranium were first undertaken by Soddy.[1] A kilogram of uranium nitrate in solution was employed. This was first chemically treated, to remove most of the small quantity of radium originally present, and was then allowed to stand in a closed vessel. The equilibrium amount of emanation formed in the solution was then tested at intervals. Preliminary experiments showed that the rate of production of the radium was certainly far slower than the theoretical value, and at first little if any indication of production of radium was observed. In later experiments, however, Soddy found that in the course of eighteen months, the amount of radium in the solution had distinctly increased.

The solution after this interval contained about 1.6×10^{-9} gram of radium. This gives the value of about 2×10^{-12} as the fraction of uranium changing per year, while the theoretical fraction is 2×10^{-10}, or 100 times greater than the observed amount.

Whetham also found a similar result, but concluded that the rate of production was faster than that observed by Soddy. On the other hand, Boltwood finds no certain evidence of the growth of radium from uranium, although an extremely minute quantity was detectable in his apparatus. In his experiments, 100 grams of uranium were obtained almost completely free from radium by fractional crystallization. After this treatment, no trace of radium could be detected in his uranium solution, although he could with certainty have detected the presence of 1.7×10^{-11} grams.

After standing for a year, no effect was produced by the emanation in his electroscope, which was of the same sensitive-

[1] Soddy: Nature, May 12, 1904, Jan. 19, 1905; Phil. Mag., June, 1905.

ness as in the first experiments. Such a result shows that uranium, when purified in the manner adopted by Boltwood, certainly does not, in the course of a year, grow a measurable quantity of radium, and that the quantity is not more than one thousandth of the theoretical amount.

Although the experimental evidence is somewhat conflicting, I think there can be little doubt that the uranium of Soddy did show a growth of radium, although only a fraction of the amount to be expected theoretically. So far as is known at present, uranium breaks up with the expulsion of an α particle and produces uranium X, which has a period of 22 days and emits only β and γ rays. No further active product has been detected, so that we are unable to say what further stages of disintegration appear before radium is formed. If, for example, the disintegration product of UrX is a rayless substance with a very slow period, the slow rate of production of radium by uranium is at once explained. Suppose, for example, that the uranium, as in the experiments of Boltwood, was carefully purified. It is probable that the rayless product would be completely removed from the uranium. Before radium could be produced at an appreciable rate, the intermediate rayless product must be formed in some quantity. If the rayless product had a period of several thousand years, an interval of several years would be required before the appearance of radium could be detected.

Such an hypothesis of an intermediate transition product would also account for the discrepancy between the experiments of Soddy and Boltwood. In the experiments of the former, the trace of radium initially observed in the uranium was partly removed by the precipitation of barium in the uranium solution. This may not have removed the intermediate product which had been collecting in the uranium for several years. Consequently, the unpurified solution used by Soddy was better suited to show the production of radium than the carefully treated solution used by Boltwood.

I think that there can be no reasonable doubt that the pure uranium solution will ultimately show the presence of radium, although an interval of several years may be required before the amount formed is detectable.

The changes occurring in uranium which lead to the production of radium are shown below.

Uranium.
↓
Uranium X.
↓
One or more unknown transition substances with long periods of transformation.
↓
Radium and its family of products.

There can be little doubt that the intermediary product or products between uranium X and radium will ultimately be separated chemically. Supposing that there is only one intermediary product, it is not unlikely that this will prove to be rayless in character. The presence of such a product could be detected by its property of producing radium initially at a constant rate. If, for example, the unknown product were completely separated from an amount of radioactive mineral which contained a kilogram of uranium, it would produce radium initially at the rate of about 4×10^{-7} gram per year, or 10^{-9} of a gram per day. This latter amount is easily measurable, and consequently a proof of the production of radium by this substance should only require observations extending over a few weeks.

The position that radium holds in regard to uranium is unique in chemistry. For the first time it is possible to predict accurately the amount of one element present when the quantity of another is known. In seems probable that such relations will ultimately be extended to include all the radioelements and their products, and possibly also some of the apparently non-radioactive substances; for it is remarkable how certain elements are always found together in mineral deposits in about the same relative amounts, although there is no apparent chemical reason for their association.

CHAPTER VII

TRANSFORMATION PRODUCTS OF URANIUM AND ACTINIUM, AND THE CONNECTION BETWEEN THE RADIOELEMENTS

We have in previous chapters analyzed in some detail the series of transformations that take place in thorium and radium. As the two other radioactive substances, uranium and actinium, are also of interest in this connection, a brief review will now be given of the changes taking place in them.

Changes in Uranium

Uranium products give out α, β, and γ rays, but no definite evidence has yet been obtained that uranium gives off an emanation. In this respect, it appears to differ from thorium, radium, and actinium. It is, however, possible that a closer investigation may yet disclose the presence of an emanation with a very short life. If an emanation were emitted which lasted for less than a hundredth part of a second, its detection by the electric method would be extremely difficult.

Only one direct transformation product, called uranium X, has so far been observed in uranium. The separation of this substance was first effected by Sir William Crookes[1] by two distinct methods. Ammonium carbonate in excess was added to a uranium solution and the uranium precipitated. A light precipitate remained behind which contained the UrX. Crookes used the photographic method and observed that the uranium, after this treatment, was photographically almost inactive, while the precipitate containing the UrX, when compared with an equal weight of uranium, had a very intense photographic action. The explanation of this was made clear by later experiments. UrX gives out only β rays, which, in the case of uranium, produce far more photographic action than the easily

[1] Crookes: Proc. Roy. Soc., lxvi, p. 409 (1900).

absorbed α rays. The removal of UrX does not in any way alter the α ray activity of uranium, measured by the electric method, but completely removes the β ray activity.

The second method used by Crookes was to dissolve uranium in ether, when the uranium divides itself unequally between the ether and water present. The water fraction contains all the UrX, while the ether fraction is photographically inactive.

Still another means of separation of UrX was used by Becquerel.[1] A small quantity of a barium salt was added to a uranium solution, and then precipitated by the addition of sulphuric acid. The dense barium precipitate carries down the UrX with it, and, after several successive treatments, the UrX is almost completely removed from the uranium. Becquerel first noted that the UrX loses its activity after some time, while the uranium recovers its lost activity.

The rate at which UrX loses its activity was determined by Rutherford and Soddy. The decay curve, like the decay curves for simple radioactive products, is exponential, and UrX loses half its activity in about 22 days. The recovery curve of uranium measured by the β rays, due to the fresh production of UrX in the uranium, is complementary to the decay curve.

From analogy with the corresponding results observed in thorium and radium, we may thus conclude that uranium produces the new product UrX at a constant rate. Since the α ray activity is unaffected by the removal of UrX, it seems probable that the uranium atom breaks up with the emission of an α particle and then becomes the atom of UrX. This in turn breaks up with the expulsion of a β particle. The product resulting from the transformation of UrX is either inactive, or active to such a feeble degree that its transformation cannot be directly followed by the electric method.

The changes taking place in uranium are diagrammatically illustrated below.

[1] Becquerel: Comptes rendus, cxxxi, p. 137 (1900); cxxxiii, p. 977 (1901).

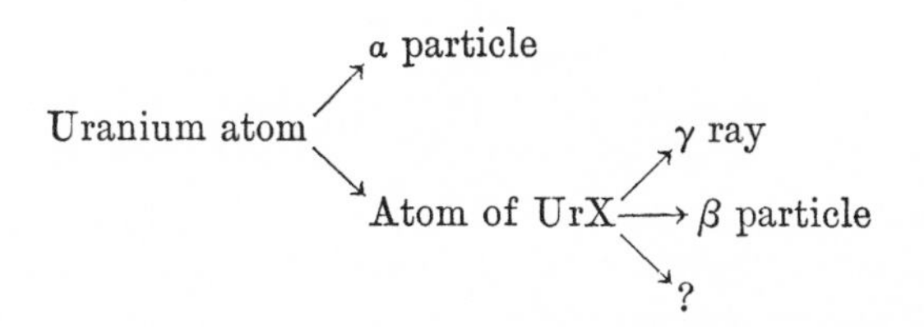

It has been pointed out in the last chapter that UrX probably undergoes one or more further changes of a long period, possibly rayless in character, and is finally transformed into radium.

There are several points of interest in connection with the β ray activity exhibited by uranium. Meyer and Schweidler[1] drew attention to some remarkable variations of the β activity of uranium during crystallization under various conditions. This activity varies in a most capricious manner, as if the process of crystallization had some direct effect on the rate of transformation of UrX. Some later experiments made by Dr. Godlewski[2] in the laboratory of the writer finally led to a simple explanation of these puzzling phenomena observed by Meyer and Schweidler.

Some uranium nitrate was heated and sufficient water added for complete solution. A small dish containing the heated solution was then placed under a β ray electroscope. The β ray activity of the solution remained sensibly constant during the cooling of the solution, but the moment crystallization commenced at the bottom of the dish, the β ray activity increased rapidly, and reached several times its initial value at the completion of the crystallization. After reaching a maximum, the activity gradually diminished again, and about a week later had reached a value equal to that of the uranium nitrate before solution.

Another simple experiment was then made. A cake of crystals so formed was removed from the dish immediately after crystallization was completed, and inverted under the electroscope. The β ray activity was much less than for the other side of the cake, and gradually increased again to the

[1] Meyer and Schweidler: Wien Ber., cxiii, July, 1904.

[2] Godlewski: Phil. Mag., July, 1905.

normal value. The explanation of this result is as follows. UrX is more soluble in water than uranium itself. When the crystallization starts at the bottom of the dish, the UrX is pushed towards the surface of the solution. The β rays entering the electroscope have on an average to pass through a less depth of uranium than before. The β ray activity will thus increase until the crystallization is complete. The lower surface of the plate of crystals will contain less than the normal amount of UrX, and consequently will show a smaller β ray effect. The gradual decrease of the β ray activity of the upper surface, and the increase of activity of the lower, appears to be due to a diffusion of the UrX through the mass of crystals. This process continues until the UrX is again uniformly distributed throughout the crystalline mass. This diffusion takes place comparatively rapidly even in a plate of completely dry crystals. An effect of this kind, which is quite likely to occur in any mixture of products differing in solubility, shows how much care is necessary in interpreting variations of activity in a mass of substance which has just been subjected to chemical treatment.

The fact that UrX is more soluble in water than uranium can be simply utilized to effect a partial separation of UrX. If uranium nitrate is dissolved in a slight excess of water, the liquid left on the surface after crystallization contains a large fraction of the total amount of the UrX originally present in the uranium.

Changes in Actinium

Shortly after the discovery of radium and polonium, Debierne noted the presence of a new radioactive substance in pitchblende residues which he called actinium. This was removed from the radioactive mineral with the thorium, but can be separated from it by suitable methods. Very little was known for several years about the radioactive peculiarities of this substance. In the meantime, Giesel had independently observed that a new radioactive substance was removed with the lanthanum and cerium present in the radioactive mineral. This substance emitted very freely a short-lived emanation, and it was for this reason that he first termed it the "emanating substance" — a name which was later changed to "emanium." Debierne found that actinium gave out an emanation which lost half its activity in 3.9 seconds. Later work by various observers has shown that the emanation and the excited activity produced by emanium and actinium have the same rates of decay.

The active constituent present in the actinium of Debierne is thus identical with that in the "emanium" of Giesel, and the original name "actinium" will in consequence be used for this substance. Actinium has not yet been separated in a sufficiently pure state to examine its atomic weight or spectrum.

Very active preparations of actinium have already been obtained by Giesel and Debierne, and it seems probable that in the pure state actinium will prove to be of the same order of activity as radium. The emanation is given out very freely from the preparations of Giesel, and excites phosphorescence on a zinc sulphide screen brought near it. The phenomenon of scintillations is shown by actinium rays to an even more marked degree than by the α rays of radium. The continuous and rapid emission of a short-lived emanation from actinium can be simply illustrated by a very striking experiment. A small quantity of actinium, enclosed in a paper envelope, is placed on a zinc sulphide screen. The α particles emitted from the mass of the substance are stopped by the paper, but the emanation readily diffuses through it into the surrounding air. The α particles expelled from the emanation produce luminosity in the zinc sulphide screen. On examination with a lens, this light is seen to be made up of a multitude of brilliant scintillations. A puff of air removes the emanation, and the luminosity disappears for a moment, but returns almost immediately as a fresh amount of emanation is supplied. The luminosity rapidly spreads from the actinium over the screen by the process of diffusion. The slightest current of air produces a marked wavering effect on the luminosity and displaces the luminosity in the direction of the air current.

Actinium gives out α, β, and γ rays. These radiations have been examined by Godlewski.[1] The β rays are apparently fairly homogeneous, and have less power of penetration than the corresponding rays from other active substances. This shows that the β particles are all projected at about the same velocity, and that this velocity is less than that of the average β rays from other substances.

The γ rays also have much less penetrating power than those from radium. It seems not unlikely that the absence of very penetrating γ rays is connected with the absence of swiftly moving β particles, for it is probable that the β particle, which is projected from radium with a velocity nearly that of light, will give rise to a more penetrating pulse than one projected at a much lower speed.

In radioactive properties, actinium shows a remarkable similarity to thorium. It emits a short-lived emanation, and this is transformed into an active deposit which is concentrated on the negative electrode in an electric field.

The activity of the deposit obtained by a long exposure to the emanation subsequently diminishes, and ten minutes after removal from the emanation, decays exponentially with a period of about 34 minutes. Miss Brooks[2] showed that the curves of excited activity for a short exposure exhibited the same general behavior as the corresponding curves obtained for the active deposit of thorium. The activity at first increased, passed through a maximum after about 8 minutes, and finally decayed exponentially, with a period of 34 minutes.

These results admit of the same explanation as in the case of the active deposit of thorium. The emanation which gives out α rays changes into a rayless product, actinium A, which is half transformed in 34 minutes. This changes into another substance called actinium B, which is half transformed in about 2 minutes, and emits α, β, and γ rays.

The choice of the 2 minute period for actinium B rather than for actinium A followed from an observation of Miss Brooks.

[1] Godlewski: Phil. Mag., Sept., 1905.

[2] Miss Brooks: Phil. Mag., Sept., 1904.

The active deposit, obtained on a platinum plate, was dissolved in hydrochloric acid. The solution was then electrolyzed, and an active substance which emitted α rays was obtained on one of the electrodes. This lost its activity exponentially with the period of about 1.5 minutes. This result shows that actinium B, which emits rays, must have the shorter period.

The analogy with thorium became still closer when Godlewski[1] and Giesel[2] independently separated from actinium a very active substance called actinium X. This was effected by precipitation with ammonia in exactly the same way as is required for the separation of ThX from thorium. The actinium X after precipitation of the actinium remains behind in the filtrate mixed with actinium A and B. Godlewski found that actinium X lost its activity exponentially with a period of about 10 days. The actinium, freed from actinium X, at the same time recovered its activity. There are, however, several interesting points of difference in the chemical separation of actinium X and ThX from their respective elements. In the case of thorium, thorium A and B are only slightly soluble in ammonia, and consequently are not removed with the ThX. Quite the reverse holds for actinium. The active deposit is readily soluble in ammonia, and consequently is separated with actinium X.

After removal of actinium X by successive precipitations, the actinium itself retains only a small proportion of its normal activity, while in the case of thorium, the residual α ray activity is about one quarter of the total. It seems probable that, if the actinium were completely freed from actinium X and its subsequent products, the element itself would show no activity measured by the α or β rays, or, in other words, that actinium itself is a rayless product. From the results of Hahn, discussed on page 168, it has already been pointed out that thorium freed from radiothorium may also prove to be a rayless substance.[3]

[1] Godlewski: Phil. Mag., July, 1905.

[2] Giesel: Jahrbuch. d. Radioaktivität, i, p. 358 (1904).

[3] Hahn (Nature, April 12, 1906) has recently separated another product from actinium which he has called "radioactinium." This product is intermediate between actinium and actinium X, emits α rays, and has a period of transformation of about

Godlewski showed that the emanation from actinium was a direct product of actinium X, and not of actinium itself. In this respect, ThX and actinium X have very similar properties. The transformations taking place in actinium are shown in Fig. 40.

On comparison of the changes taking place in actinium and thorium (see Fig. 41) the similarity in the succession of changes in the two substances is very noteworthy. Not only are the products equal in number, but the corresponding products are closely allied in general chemical and physical properties. The active deposit of actinium differs somewhat from that of thorium

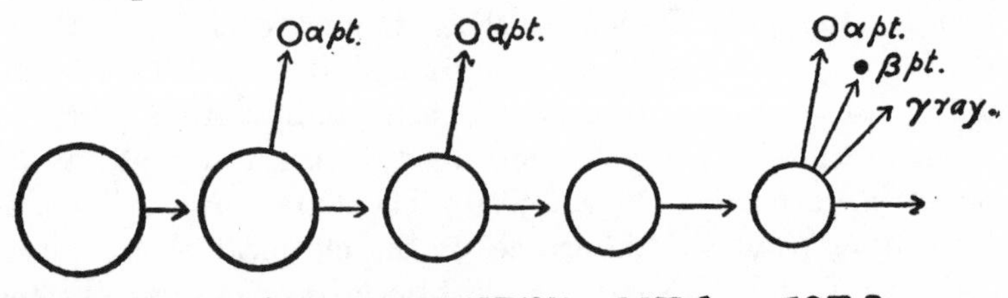

FIG. 40.

Actinium and its family of products.

in the ease with which it is dissolved by various solutions and the lower temperature at which it is volatilized.

This similarity in the radioactive changes of the two substances indicates that the atoms of actinium and thorium, while chemically distinct, are very similarly constituted, and that when once the process of disintegration is started, the atom of both substances passes through a similar succession of changes.

CONNECTION BETWEEN THE RADIOELEMENTS

The series of transformations taking place in the radioelements are shown in Fig. 41.[1]

20 days. Actinium itself is a rayless product. Godlewski had unknowingly separated this product from his actinium, for otherwise the actinium would have emitted α rays, due to the presence of radioactinium. Levin has found that actinium X does not emit β rays. The β rays from actinium arise only from the product actinium B.

[1] In the diagram (Fig. 41), the products radiothorium, radioactinium, should be introduced between thorium and thorium X, actinium and actinium X, respectively.

The substances thorium, radium, and actinium exhibit many interesting points of similarity in the course of their transformation. Each gives rise to an emanation whose life is short compared with that of the primary element itself. Such experiments as have yet been made, indicate that these emanations have no definite combining properties, but belong apparently

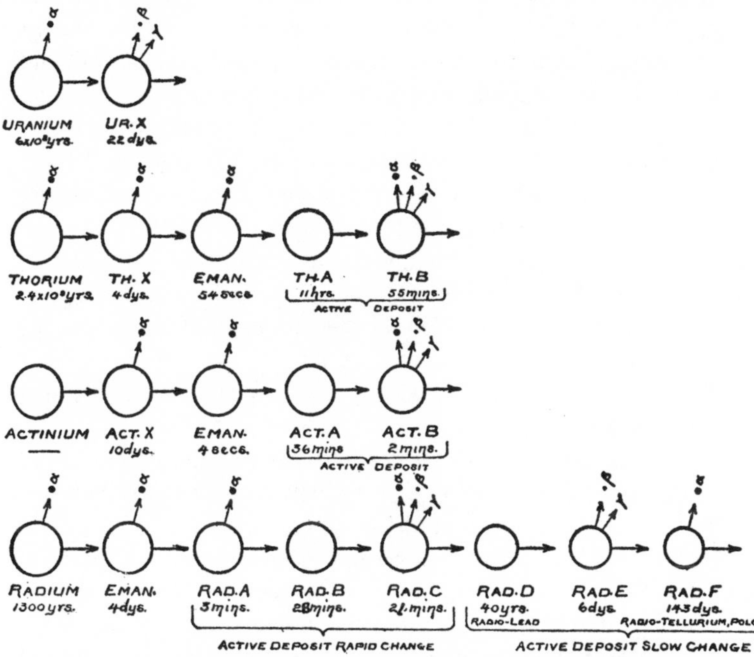

Fig. 41.

The radioelements and their family of products.

to the helium-argon group of inert gases. In each case, the emanation gives rise to a non-volatile substance which is deposited on the surface of bodies and is concentrated on the negative electrode in an electric field. The changes in these active deposits are also very similar, for each gives rise to a rayless product, followed by a product which emits all three types of rays. In each case, also, the rayless product has a longer period, or, in other words, is a more stable substance than the ray product which results from its transformation.

The disintegration of the corresponding products thorium B, actinium B, and radium C is of a more violent character than is observed in the other products, for not only is an α particle expelled at a greater speed, but a β particle is also thrown off at great velocity. After this violent explosion within the atom, the resulting atomic system sinks into a more permanent state of equilibrium, for the succeeding products thorium C and actinium C have not so far been detected by radioactive methods, while radium D is transformed at a very slow rate.

This similarity in the properties of the various families of products is too marked to be considered a mere coincidence, and indicates that there is some underlying law which governs the successive stages of the disintegration of all the radioelements. The transformation products mark the distinct stages in the career of disintegration of the atoms, and represent the halting places where the atoms are able to exist for an appreciable time before again breaking up into other more or less stable configurations.

The interesting question arises whether the atom after losing an α particle is able to exist for a short time in more than one stable form. After the expulsion of an α particle with explosive violence, there must result a rearrangement of the parts of the atom to form a permanently or temporarily stable system. It is conceivable that more than one fairly stable arrangement may be possible, and, in such a case, two or more products of disintegration must be produced in addition to the expelled α particles. These stable atomic systems, although of equal atomic weights, would exhibit differences in chemical properties, and it should be possible to separate them from one another. It is not necessary that these products should be formed in equal amount. One might exist in comparatively large amount compared with the others.

There is in addition another possibility to be borne in mind. The violent disturbance in the atom resulting in the expulsion of an α particle may cause an actual breaking up of the main atom into two parts, and thus give rise to an equal number of atoms of different atomic weights in addition to the α particle. For example, such an effect might arise during the violent disintegration of radium C or thorium B.

So far, it has not been found necessary to choose between these theories to explain the transformation products of the different elements. The disintegration in each case results in the appearance of only one substance in addition to the expelled particles. It is not unlikely, however, that a still closer examination of the radioelements may show the existence of products which lie outside the main line of descent. The method of electrolysis has already proved of great value in separating products of the transformation of radioelements which are present in infinitesimal amount in a solution, and its possibilities in this direction are by no means exhausted.

Rayless Transformations

We have seen that the great majority of the products break up with the expulsion of an α particle; in addition, a small number emit a β particle with its accompaniment the γ rays, while a few emit only a β particle. There is also a special class of product which does not emit rays at all.

It has been shown that two of these rayless products exist in radium and actinium, and probably two in thorium. The method of showing the existence of such rayless products and of determining their physical and chemical properties has already been discussed in previous chapters. Since a rayless product does not emit any ionizing type of radiation, its presence can only be observed indirectly by examination of the variation in the amount of the succeeding product. By such methods, we are enabled not only to determine the period of change of the rayless product, but also its more marked chemical and physical properties.

These products are apparently similar in all respects to the ray products, with the exception that there is no evidence of the emission of α or β particles. They are unstable substances which break up according to the same law as the other active products, and give rise to another substance of different physical and chemical properties.

There are two general ways of regarding the transformation of a rayless product. In the first place, it may be supposed that the transformation consists, not in an actual expulsion of a part of the atomic system, but in a rearrangement of the component parts of the atom to form a new temporarily stable system. On such a view, the atom of the rayless product has the same atomic weight as the succeeding product, but differs from it so materially in atomic configuration that the physical and chemical properties are quite distinct. The two products may thus be considered to be somewhat analogous to the case of an element like sulphur, which exists in two distinct forms. This analogy is, however, only superficial, for the atoms of the products possess entirely different chemical and physical properties whether in the solid state or in solution.

On the other hypothesis, the transformation of a rayless product is supposed to be similar in character to that of a ray product, the only difference being that the α particle is not expelled with sufficient velocity to produce appreciable ionization of the gas. There is an actual loss of mass during the transformation, but this loss cannot be detected by the electric method. In the light of some experimental results, discussed in Chapter X, such an explanation appears not improbable. It is there shown that when the velocity of the α particle falls below about 40 per cent of the maximum velocity of the swiftest α particle from radium, viz., that expelled from radium C, the photographic, phosphorescent, and ionizing properties of the α particle become relatively very small. Since the α particle from radium C is projected with a velocity of about 1/15 the velocity of light, it is seen that an α particle may be projected from matter at a great speed, and yet produce a comparatively weak electrical effect compared with that produced by a particle projected at twice its velocity. The α particle from radium C produces about 100,000 ions in the gas before it is absorbed, and consequently the electrical effect due to the charged α particles alone would be insignificant in comparison with that due to the ionization of the gas by the passage of the swift α particle through it. Remembering that a rayless product is generally followed by a product which emits high velocity α particles, the strong ionization effect due to the latter would tend to mask completely the small electrical effect due to the rayless product alone, even if it emitted charged α particles at slow velocity.

It is difficult to devise experiments to decide which of these two hypotheses as to the nature of a rayless transformation is correct, but the view that there is an expulsion of an α particle at too low a velocity to be detected by ordinary means has many points in its favor.

Properties of the Products

We have seen that, with a few exceptions, the products of transformation of the radioelements exist in too small quantities to be ever detected by direct measurement of their weight or volume. Even though the products exist in infinitesimal amount in the parent matter, the property of emitting ionizing radiations serves not only to measure their rate of transformation, but also to deduce some of their physical and chemical properties.

The electric method has been utilized as an accurate means of qualitative and quantitative analysis of radioactive matter which is present in extraordinarily small amount. The presence of 10^{-11} gram of a slowly changing substance like radium can easily be observed, while in the case of more rapidly changing matter like the thorium emanation, one hundred millionth of this small amount is readily detectable. In fact, as has been previously pointed out, the electric method is easily capable of showing the presence of radioactive matter in which only one atom breaks up per second, provided that a high velocity particle is expelled during the transformation.

With the aid of the electroscope, the range of possible application of chemical methods of separation has been enormously extended. It has been found that the ordinary methods of chemical separation of substances, whether depending on differences in solubility or volatility, or upon electrolysis, still apply to matter existing in infinitesimal proportion. For the detection of minute amounts of active matter, the electroscope far transcends in delicacy the balance or even the spectroscope.

The study of radioactivity has thus indirectly furnished chemistry with new methods of attack on the properties of active matter existing in extremely small quantity. Much still remains to be done in this new field of work whose importance is not as yet sufficiently recognized.

Attention has already been drawn to the radical alteration in properties of successive transformation products. This is well exemplified by the transformation of radium into its emanation and of the emanation into the active deposit. Each of these substances is entirely dissimilar in physical and chemical nature to the others, and, but for other evidence, it would be difficult to believe that these substances were derived from the direct transformation of the radium atom.

The atom at most stages of its disintegration loses an α particle, which has an apparent mass about twice that of the hydrogen atom. This decrease in the mass of the atom of about one per cent gives rise to an entirely new atomic configuration whose chemical and physical properties bear, as we have seen, no obvious relation to the parent atom. This radical change in the properties of the atom is not, however, very surprising if we consider chemical analogies. Elements which do not differ much in atomic weight often possess entirely dissimilar properties, and thus we might reasonably expect that a decrease of the atomic mass would result in a marked change of the chemical and physical nature of the substance.

There cannot now be any doubt that the radioactive products arise from the successive transformations of the atoms of matter, and not of the molecules. Each transformation product is a distinct element, which differs only from the well known inactive elements in the comparative instability of the atoms composing it. There can be no doubt, for example, that the radium emanation, while it lasts, is a new elementary substance with an atomic weight and spectrum that distinguish it from all other elements. If, for instance, it were possible to examine chemically any one of the simple products in a time which is short compared with its period of transformation, the substance would be found to have all the distinctive properties of a new element. It would possess a definite atomic weight and spectrum and other distinctive physical and chemical properties. In regard to their position as elements, no line of demarcation can be drawn between the comparatively stable elements like uranium, thorium, and radium, and their rapidly changing products. From a radioactive point of view, the atoms of these substances differ from one another mainly in stability. The atoms of each radio-element may differ enormously in stability, but ultimately, if sufficient time is allowed, all of these substances must be transformed through a succession of stages and disappear. There will finally remain only the inactive or stable products of their decomposition.

There is no evidence that the process of disintegration, when once started, is reversible under ordinary conditions. We can obtain the radium emanation from radium, but cannot change the emanation back again into radium. The question whether this process has been reversible under some possible conditions existing during the earth's history will be considered later in Chapter IX.

Life of the Radioelements

We have seen that every simple product which emits a radiation decreases in amount on account of its transformation into another substance. The rate of transformation is directly porportional to the constant λ, and inversely proportional to the period of transformation. The period of transformation of any simple product may be taken as a comparative measure of the stability of the atoms composing it. It is at once seen that the atomic stability of the products whose rate of transformation has been directly measured varies over an enormous range. For example, the atoms of radium F, which are half transformed in 140 days, are over three million times as stable as the atoms of the actinium emanation which is half transformed in 3.9 seconds.

This range of stability of the atoms is still further extended when we include the atoms of the primary elements uranium and thorium.

The periods of transformation of these substances can be approximately deduced by comparison of their α ray activities. Since uranium is the parent of radium, the relative amounts of uranium and radium present in an old radioactive mineral are directly proportional to the periods of transformation of the two substances. Now it has been shown that 3.8×10^{-7} gram of radium is present per gram of uranium in any radioactive mineral. Since radium is half transformed in about 1300 years, uranium must be half transformed in $1300 \times \frac{10^7}{3.8}$, or about 3.4×10^9 years.

The period of transformation of thorium is probably three or four times greater than this, since its activity is about the same as that of uranium, but gives rise to four α ray products to the one of uranium. In order that a large fraction of any given mass of uranium may be transformed, a period of at least ten thousand million years is necessary.

The period of transformation of actinium cannot be determined until it has been obtained in a pure state. If, however, its activity is of the same order as that of radium, its period will also be of the same order.

There appears to be no obvious relation existing between the periods of the successive products nor between the periods of the different families of products. It is a matter of remark, however, that a substance of great stability is generally followed by a number of comparatively unstable products. This is well exemplified in the case of thorium, radium, and actinium, where most of their known products suffer rapid transformation.

Connection between Uranium, Radium, and Actinium

The connection that exists between uranium and radium, and its products, radiolead and polonium, has been clearly brought out, and it is of interest to examine whether any similar relation exists between uranium and thorium, and uranium and actinium. The latter substance is always found in uranium minerals, and since it probably has a radioactive life comparable with that of radium, it must be produced in some way from the parent substance uranium.

This question was examined by Dr. Boltwood and the writer. If actinium, for example, was a product of uranium in the main line of descent, the activity due to actinium or to a uranium mineral should be comparable with that of radium. Since there would be a state of equilibrium between the uranium and actinium, the same number of atoms of each should break up per second. Since actinium has four a ray products and radium five, the activity due to actinium in the mineral should be comparable with that due to radium and its family of products. Experiment, however, showed that the activity observed in Colorado uraninite, for example, was almost entirely due to the uranium and radium contained in it. Its activity due to actinium was certainly only a small portion of that due to radium and its products. It seems probable that actinium does arise from uranium, but that it is not a lineal descendant of uranium in the same sense that radium is. It has already been pointed out that in some of the transformations, two distinct transition substances may be produced. It appears likely that actinium will prove to be derived from uranium or one of its products, but that it is produced in much less amount than the other product. Such a relation would explain the connection that apparently exists between uranium and actinium, and at the same time would account for the small amount of actinium present.

In regard to the connection between thorium and uranium, the evidence is not very definite. Many minerals contain uranium and very little thorium, but Strutt has shown that every thorium mineral examined contains some uranium and radium. Strutt has suggested that thorium is the parent of uranium. Such a relation is suggested by an analysis of the mineral thorianite. This mineral, of very great geological age, is found in Ceylon, and contains about 70 per cent of thorium and 12 per cent of uranium. The uranium in this mineral may have been derived from the decomposition of the thorium. There is, however, a serious objection to this view, for the atomic weight of thorium, 232.5, is less than that of the usually accepted atomic weight of uranium, 238.5. If these atomic weights are correct, it does not appear likely that thorium is the parent of uranium, unless the process of production of uranium from thorium is very different from that usually observed in radioactive transformations.

CHAPTER VIII

THE PRODUCTION OF HELIUM FROM RADIUM AND THE TRANSFORMATION OF MATTER

THE history of the discovery of helium possesses some features of unusual dramatic interest. In 1868, Janssen and Lockyer observed in the spectrum of the sun's chromosphere a bright yellow line, which could not be identified with that of any known terrestrial substance. Lockyer gave the name "helium" to this supposed new element. Further comparison showed that certain other spectral lines in the chromosphere always accompanied the yellow line and were probably characteristic of helium.

The spectrum of helium is observed not only in the sun, but also in many of the stars; and in some classes of stars, now known as helium stars, the spectrum of helium predominates. No evidence of the existence of helium on the earth was discovered until 1895. Shortly after the discovery of argon in the atmosphere, by Lord Rayleigh and Sir William Ramsay, a search was made to see if argon could be obtained from mineral sources. In 1895, Miers in a letter to "Nature" drew attention to some results obtained by Hillebrande of the U. S. Geological Survey in 1891. In the course of the detailed analysis of many of the minerals containing uranium, a considerable quantity of gas was found by Hillebrande[1] to be given off on solution of the minerals. At the time he thought this gas was nitrogen, although attention was drawn to some peculiarities of its behavior as compared with ordinary nitrogen. The mineral clevite especially gave off a large quantity of gas when heated or dissolved. Ramsay procured some of this mineral in order to see whether this gas might prove to be argon. On introducing the

[1] Hillebrande, Bull. U. S. Geolog. Survey, No. 78, p. 43 (1891).

gas liberated from clevite into a vacuum tube, a spectrum was observed entirely different from that of argon.[1] The spectrum was carefully examined by Lockyer[2] and found to be identical with that of the new element helium, previously discovered by him in the sun. After a lapse of thirty years since its discovery in the sun, helium had at last been found to exist in the earth. An examination of the properties of helium soon followed. It has a well-marked complex spectrum of bright lines, of which the most noticeable is a bright yellow line D_3 close to the sodium D lines.

It is a light gas about twice as dense as hydrogen, and, excepting the latter, has a lighter atom than any other known element. Like argon, it refuses to combine with any other substance, and must therefore be classed with the group of chemically inert gases discovered by Ramsay in the atmosphere. By measurement of the velocity of sound in a tube filled with helium, the ratio of its two specific heats was found to be 1.66. The ratio for diatomic gases like hydrogen and oxygen is 1.41. This result suggests that helium is monatomic, *i.e.*, that the helium molecule consists of only one atom; or, in other words, that the atom and molecule of helium are the same. Since the density of helium was found to be 1.98 times that of hydrogen at the same temperature and pressure, and since the hydrogen molecule contains two atoms, it was concluded that the atomic weight of helium is twice this amount, or 3.96. It must be remembered that this atomic weight has been determined only from density observations, since helium cannot be made to enter into any chemical combination, and consequently the value given for its atomic weight has not the same claim to accuracy as the atomic weights of many of the other elements which have been determined by more rigorous chemical methods.

Helium was found to exist in minute proportion in the atmosphere. In a recent paper Ramsay has concluded that one volume of helium is present in 245,000 volumes of air. The occurrence of helium in certain minerals was most remarkable, for there ap-

[1] Ramsay, Proc. Roy. Soc., lviii, p. 65 (1895).

[2] Lockyer, Proc. Roy. Soc., lviii, p. 67 (1895).

peared no obvious reason why an inert gaseous element should be found associated with minerals, which in many cases are impervious to the passage of water or gases.

Quite a new light was thrown on this subject as a result of the discovery of radioactivity. On the disintegration theory of radioactivity, it was to be expected that the final or inactive products of the transformation of the radioelements would be found in the radioactive minerals. Since many of the radioactive minerals are of extreme antiquity, it was reasonable to suppose that the inactive products of the transformation of radioactive substances, provided they did not escape, would be found associated in some quantity with the radioactive matter as its invariable companions. In looking for a possible disintegration product, the occurrence of helium in all radioactive minerals was noteworthy, for helium is mainly found in minerals which contain a large quantity of uranium or of thorium.

For these and other reasons, Rutherford and Soddy[1] suggested that helium might prove to be a disintegration product of the radioelements. Additional weight was lent to this suggestion by the writer's discovery that the α particle expelled from radium has an apparent mass about twice that of the hydrogen atom and might prove to be an atom of helium.

In the beginning of 1903, thanks to Dr. Giesel of Braunschweig, small quantities of pure radium bromide were placed on the market. Ramsay and Soddy obtained 30 milligrams of the bromide and proceeded to see if it were possible to detect the presence of helium in the gases released from it. In the first experiment the radium bromide was dissolved in water and the accumulated gases drawn off. It was known that radium bromide produced hydrogen and oxygen, and these gases were removed by suitable methods. A small bubble of residual gas was obtained which, on introduction into a vacuum tube, showed the characteristic D_3 line of helium.[2] Using another and somewhat older sample of radium bromide, lent by the writer, the

[1] Rutherford and Soddy: Phil. Mag., p. 582, 1902, pp. 453 and 579, 1903.

[2] Ramsay and Soddy: Nature, July 16, p. 246, 1903. Proc. Roy. Soc., lxxii, p. 204 (1903); lxxiii, p. 346 (1904).

residual gas released by solution of the radium was found to give a complete spectrum of helium.

This experiment showed that helium was produced by radium and retained to some extent in the solid compound. Further experiments revealed a still more interesting fact. The emanation from the 60 milligrams of radium bromide was condensed in a glass tube and the other gases pumped out. After volatilization, the emanation was introduced into a small vacuum tube. The spectrum at first showed no sign of the helium lines, but after three days the D_3 line of helium made its appearance, and after five days the complete spectrum of helium was observed. This experiment shows that helium is produced from the emanation, for no evidence of its presence was obtained immediately after the introduction of the emanation into the spectrum tube.

The discovery of the production of helium by the radium emanation was of great importance, as it showed in a striking manner the extraordinary nature of the processes occurring in radium, and was the first definite evidence of the possibility of one element being transformed into another stable element. The experiments were not easy of performance, as the helium was present only in minute amount, and the experience gained by Ramsay in his previous work on the rare gases in the atmosphere was of the greatest practical value in bringing the experiments to such a successful conclusion.

The production of helium by radium has been confirmed by a number of experimenters. Curie and Dewar[1] made in this connection a most interesting experiment, which showed conclusively that the helium was produced from radium and could not be ascribed to a possible occlusion of helium in the radium bromide. A large quantity of radium chloride was introduced into a quartz tube and the radium heated to fusion. The emanation and gases in the tube were pumped out and the tube sealed. One month later, Deslandres examined the spectrum of the gases in the tube by placing layers of foil over the ends. A complete spectrum of helium was observed,

[1] Curie and Dewar: Comptes rendus, cxxxviii, p. 190 (1904).

showing that the helium had been produced from the radium in the interval.

Recently Debierne[1] has found that helium is also produced from active preparations of actinium. This result shows that the helium must be a common product of these two substances, which, from their radioactive and chemical behavior, must be regarded as distinct elements.

The Position of Helium as a Transformation Product of Radium

We have already seen that radium is transformed through a long succession of products, each of which has some distinctive physical and chemical properties and a definite period of transformation. These products differ from the ordinary chemical elements only in the instability of their component atoms. They must be regarded as transition elements with a limited life, which break up into new forms of matter at a rate independent of our control.

The distinction, however, between helium as a product of radium and the family of transition products is mainly one of atomic stability. As far as we know, helium is a stable element which does not disappear, but in the case of all the radioactive products, including the primary sources uranium and thorium, the atoms are undoubtedly unstable.

It is now necessary to consider the position of helium as a transformation product of radium. Some have considered that helium is the end or final product of the disintegration of the radium atom, but for this there is no experimental evidence. We have seen that after the first rapid changes in the active deposit of the emanation, there is produced a very slow transition substance, radium D. If helium were the final product of the transformation of the radium atom, the amount produced from the emanation in the course of a few days would be infinitesimally small. In addition there can be little doubt that the final active product of radium, viz. radium F (polonium), is an element of high atomic weight.

[1] Debierne: Comptes rendus, cxli, p. 383 (1905).

The evidence, on the other hand, points strongly to the conclusion that the helium is formed by the α particles continuously shot out from radium and its products. We shall see later (Chapter X) that the experimental evidence shows that the α particle shot out from the different α ray products of radium has in each case the same mass, but varies in velocity for the different products.

From observation of the deflection of the rays both in a strong magnetic and a strong electric field, the velocity, and the value e/m — the ratio of the charge of the α particle to its mass — have been accurately measured. The ratio e/m is very nearly 5×10^3. Now the ratio e/m for the hydrogen atom liberated in the electrolysis of water is known to be 10^4. If we assume that the α particle carries the same charge as the hydrogen atom, the mass of the α particle is twice that of the hydrogen atom. We are here unfortunately confronted with several possibilities between which it is at present difficult to make a definite choice.

The value of e/m for the α particle may equally well be explained on the assumptions that the α particle is (1) a molecule of hydrogen, (2) a helium molecule carrying *twice* the charge of the hydrogen atom, or (3) one half of the helium molecule carrying the usual ionic charge.

The hypothesis that the α particle is a molecule of hydrogen seems for many reasons improbable. If hydrogen is a constituent of the atoms of radioactive matter, it is to be expected that it would be expelled in the atomic and not in the molecular state. In all cases so far examined, when hydrogen is the carrier of an electric charge, the value of e/m is 10^4. This is the value to be expected for the hydrogen atom. For example, Wien found that the maximum value of e/m for the canal rays or positive ions, which are produced in an exhausted vacuum tube, was 10^4. In addition, it seems improbable that, even if the hydrogen were projected initially in the molecular state, it would escape decomposition into its component atoms in passing through matter.

When it is remembered that an α particle is projected with a velocity of about 12,000 miles per second, and collides with every molecule in its path, the disturbance set up in the molecule by the collisions must be very intense, and must tend to rupture the bonds that hold the atoms of the molecule together. Indeed, it seems very unlikely that the hydrogen molecule under such conditions could escape decomposition into its component atoms. If the *a* particle were a hydrogen molecule, a considerable amount of free hydrogen should be present in old radioactive minerals which are sufficiently dense to prevent its escape. This does not appear to be the case, although in some minerals there is a considerable quantity of water. On the other hand, the comparatively large amount of helium present supports the view that the *a* particle is connected with helium. A strong argument in favor of the view of a connection between helium and the *a* particle rests on the observed facts that helium is produced by actinium as well as by radium. The only point of similarity between these two substances lies in the expulsion of *a* particles. The production of helium by both substances is at once obvious if the helium is derived from the accumulated *a* particles, but is difficult to explain on any other hypothesis. We are thus reduced to the view that either the *a* particle is a helium atom carrying twice the ionic charge, or that it is half of a helium atom carrying an ionic charge.

The latter assumption involves the conception that helium, while behaving as a chemical atom under ordinary chemical and physical conditions, may exist in a still more elementary state as a component of the atoms of radioactive matter, and that, after expulsion, the *a* particles lose their charge and recombine to form atoms of helium.

While such a view cannot be dismissed as inherently improbable, there is no direct evidence in its favor. On the other hand, the second hypothesis has the merit of greater simplicity and probability.

On this view, the *a* particle is in reality an atom of helium which is either expelled with a double ionic charge or acquires this charge in its passage through matter. Even if the *a* particle were initially projected without a charge, it would almost certainly acquire one after the first few collisions with the molecules in its path. We know that the *a* particle is a very efficient ionizer, and there is every reason to suppose that it would itself be ionized by its collision with the molecules in its path, *i. e.*, it would lose an electron, and would consequently itself retain a positive charge.

If the *a* particle can remain stable with the loss of two electrons, these electrons would almost certainly be ejected as a result of the intense disturbance arising from the collision of the *a* particle with the molecules in its path. The *a* particle would then have twice the ordinary ionic charge, and the value of e/m, as found by measurement, would be quite consistent with the view that the *a* particle is an atom of helium.

If this be the case, the actual number of *a* particles expelled from radium would be only one half of that deduced on the assumption that the *a* particle carries a single charge. This would make the rate of disintegration of radium only half of that calculated in Chapter VI, and would consequently double its life.

In a similar way this assumption would reduce the calculated volume of the emanation released from one gram of radium from .8 c. mms. to .4 c. mms. This is smaller than the experimental value — about 1 c. mm. — determined by Ramsay and Soddy, but is of the right order of magnitude.

On the above assumptions, the volume of helium produced per year per gram of radium can readily be calculated. If each *a* particle carries twice the ionic charge, experiment shows that 1.25×10^{11} *a* particles are expelled per second from one gram of radium in equilibrium. The number expelled per year is 4.0×10^{18}. Since one cubic centimetre of a gas at standard pressure and temperature contains 3.6×10^{19} molecules, the volume of helium produced per year per gram of radium is .11 c. cms.

Ramsay and Soddy made an estimate of the rate of production of helium from radium in the following manner. The helium produced from 50 mgrs. of radium bromide kept in a closed vessel for 60 days was introduced into a vacuum tube. Another similar tube was placed in series with it and the amount of helium in the latter was adjusted until the discharge, passed in series through the two tubes, showed the helium lines with about the same intensity. In this way, they deduced the volume obtained from the radium to be 0.1 cubic mm. This corresponds to a rate of production of helium per gram of radium per year of about 20 cubic mms. This is only about one fifth of the theoretical amount calculated above. Ramsay and Soddy do not lay much stress on the accuracy of their estimate, as they consider that the presence of a trace of argon may have seriously interfered with the correctness of the estimate by the spectroscopic method. An accurate measurement of the rate of production of helium by radium would be of the utmost value at the present time in settling the connection between the α particle and helium.

If the α particle is a helium atom, the greater proportion of the α particles expelled from the emanation enclosed in a small tube will be projected into the glass envelope. The swiftest moving particles, viz., those expelled from radium C, would probably penetrate the glass to a depth of about 1/20 of a millimetre, while the slower moving particles would be stopped after traversing a somewhat shorter distance.

It has already been pointed out (page 88) that this may explain why the volume of the emanation in the first experiment by Ramsay and Soddy shrank almost to zero. The helium in this case was retained in the glass. In the second experiment the helium may have diffused from the glass tube into the gas again. Ramsay and Soddy endeavored to settle this point by testing whether helium was released by heating a glass tube in which the emanation had been enclosed for several days and then removed. The spectroscope momentarily showed some of the helium lines, but these were soon obscured by the presence of other gases liberated by the heating of the tube.

Age of Radioactive Minerals

The helium observed in the radioactive minerals is almost certainly due to its production from the radium and other radioactive substances contained therein. If the rate of production of helium from known weights of the different radio-elements were experimentally known, it should thus be possible to determine the interval required for the production of the amount of helium observed in radioactive minerals, or, in other words, to determine the age of the mineral. This deduction is based on the assumption that some of the denser and more compact of the radioactive minerals are able to retain indefinitely a large proportion of the helium imprisoned in their mass. In many cases the minerals are not compact but porous, and under such conditions most of the helium will escape from its mass. Even supposing that some of the helium has been lost from the denser minerals, we should be able to fix with some certainty a minimum limit for the age of the mineral.

In the absence of definite experimental data on the rates of production of helium by the different radioelements, the deductions are of necessity somewhat uncertain, but will nevertheless serve to fix the probable order of the ages of the radioactive minerals.

It has already been pointed out that all the α particles expelled from radium have the same mass. In addition it has been experimentally found that the α particle from thorium B has the same mass as the α particle from radium. This would suggest that the α particles projected from all radioactive substances have the same mass, and thus consist of the same kind of matter. If the α particle is a helium atom, the amount of helium produced per year by a known quantity of radioactive matter can readily be deduced on these assumptions.

The number of products which expel α particles are now well known for radium, thorium, and actinium. Including radium F, radium has five α ray products, thorium five, and actinium four. With regard to uranium itself, there is not the same certainty, for only one product, UrX, which emits only β rays, has so far been chemically isolated from uranium. The α particles apparently are emitted by the element uranium itself; at the same time, there is some indirect evidence in support of the view that uranium contains three α ray products. For the purpose of calculation, we shall, however, assume that in uranium and radium in equilibrium, one α particle is expelled from the uranium for five from the radium.

Let us now consider an old uranium mineral which contains one gram of uranium, and which has not allowed any of the products of its decomposition to escape. The uranium and radium are in radioactive equilibrium and 3.8×10^{-7} grams of radium are present. For one α particle emitted by the uranium, five are emitted by the radium and its products, including radium F. Now we have shown that radium with its four α ray products probably produces .11 c.c. of helium per gram per year. The rate of production of helium by the uranium and radium in the mineral will consequently be $\frac{5}{4} \times .11 \times 3.8 \times 10^{-7} = 5.2 \times 10^{-8}$ c.c. per year per gram of uranium.

Now, as an example of the method of calculation, let us consider the mineral fergusonite which was found by Ramsay and Travers to evolve 1.81 c.c. of helium per gram. The fergusonite contains about 7 per cent of uranium. The amount of helium contained in the mineral per gram of uranium is consequently 26 c.c.

Since the rate of production of helium per gram of uranium and its radium products is 5.2×10^{-8} c.c. per year, the age of the mineral must be at least $26 \div 5.2 \times 10^{-8}$ years or 500 million years. This, as we have pointed out, is a minimum estimate, for some of the helium has probably escaped.

We have assumed in this calculation that the amount of uranium and radium present in the mineral remains sensibly constant over this interval. This is approximately the case, for the parent element uranium probably requires about 1000 million years to be half transformed.

As another example, let us take a uranium mineral obtained from Glastonbury, Connecticut, which was analyzed by Hillebrande. This mineral was very compact and of high density, 9.62. It contained 76 per cent of uranium and 2.41 per cent of nitrogen. This nitrogen was almost certainly helium, and dividing by seven to reduce to helium this gives the percentage of helium as 0.344. This corresponds to 19 c.c. of helium per gram of the mineral, or 25 c.c. per gram of uranium in the mineral. Using the same data as before, the age of the mineral must be certainly not less than 500 million years. Some of the uranium and thorium minerals do not contain much helium. Some are porous, and must allow the helium to escape readily.

A considerable quantity of helium is, however, nearly always found in the compact primary radioactive minerals, which from geologic data are undoubtedly of great antiquity.

Hillebrande made a very extensive analysis of a number of samples of minerals from Norway, North Carolina, and Connecticut, which were mostly compact primary minerals, and noted that a striking relation existed between the proportion of uranium and of nitrogen (helium) that they contained. This relation is referred to in the following words:—

"Throughout the whole list of analyses in which nitrogen (helium) has been estimated, the most striking feature is the apparent relation between it and the UO_2. This is especially marked in the table of Norwegian uraninites, recalculated from which the rule might almost be formulated that, given either nitrogen or UO_2, the other can be found by simple calculation. The same ratio is not found in the Connecticut varieties, but if the determination of nitrogen in the Branchville mineral is to be depended on, the rule still holds that the higher the UO_2 the higher likewise is the nitrogen. The Colorado and North Carolina minerals are exceptions, but it should be borne in mind that the former is amorphous, like the Bohemian, and possesses the further similarity of containing no thoria, although zirconia may take its place, and the North Carolina mineral is so much altered that its original condition is unknown."

Very little helium, however, is found in the secondary radioactive minerals, *i. e.*, minerals which have been formed as a result of the decomposition of the primary minerals. These minerals, as Boltwood has pointed out, are undoubtedly in many cases of far more recent formation than the primary minerals, and consequently it is not to be expected that they should contain as much helium. One of the most interesting deposits of a secondary uraninite is found at Joachimsthal in Bohemia, from which most of our present supply of radium has been obtained. This is rich in uranium, but contains very little helium.

When the data required for these calculations are known with more definiteness, the presence of helium in radioactive minerals will in special cases prove a most valuable method of computing their probable age, and indirectly the probable age of the geological deposits in which the minerals are found. Indeed, it appears probable that it will prove one of the most reliable methods of determining the age of the various geological formations.

Significance of the Presence of Lead in Radioactive Minerals

If the α particle is a helium atom, the atomic weights of the successive α ray products of radium must differ by equal steps of four units. Now we have seen that uranium itself probably contains three α ray products. Since the atomic weight of uranium is 238.5, the atomic weight of the residue of the uranium after the expulsion of three α particles would be $238.5 - 12, = 226.5$. This is very close to the atomic weight of radium 225, which we have seen is produced from uranium. Now radium emits five α ray products altogether, and the atomic weight of the end product of radium should be $238.5 - 32, = 206.5$. This is very close to the atomic weight of lead, 206.9. This calculation suggests that lead may prove to be the final product of the decomposition of radium, and this suggestion is strongly supported by the observed fact that lead is always found associated with the radioactive minerals, and especially in those primary minerals which are rich in uranium.

The possible significance of the presence of lead in radioactive minerals was first noted by Boltwood,[1] who has collected a large amount of data bearing on this question.

The following table shows the collected results of an analysis of different primary minerals made by Hillebrande: —

Locality.	Percentage of uranium.	Percentage of lead.	Percentage of nitrogen.
Glastonbury, Connecticut	70–72	3.07–3.26	2.41
Branchville, Connecticut	74–75	4.35	2.63
North Carolina	77	4.20–4.53	
Norway	56–66	7.62–13.87	1.03–1.28
Canada	65	10.49	0.86

[1] Boltwood: Phil. Mag., April, 1905; Amer. Journ. Science, Oct., 1905.

Five samples were taken of the minerals from Glastonbury, three from Branchville, two from North Carolina, seven from Norway, and one from Canada. In minerals obtained from the same locality, there is a comparatively close agreement between the amounts of lead contained in them. If helium and lead are both products of the decomposition of the uranium radium minerals, there should exist a constant ratio between the percentage of lead and helium in the minerals. The percentage of helium is obtained from the above table by dividing the nitrogen percentage by seven. Since probably eight α particles are emitted from the decomposition of uranium and radium for the production of one atom of lead, the weight of helium formed should be $\frac{8 \times 4}{206.9} = .155$ of the weight of lead. This is based on the assumption that all the helium formed is imprisoned in the minerals. The ratio actually found is about .11 for the Glastonbury minerals, .09 for the Branchville minerals, and about .016 for the Norway minerals. It will be noted that in all cases the ratio of helium to lead is less than the theoretical ratio, indicating that in some cases a large proportion of the helium formed in the mineral has escaped. In the case of the Glastonbury minerals, the observed ratio is in good agreement with theory.

If the production of lead from radium is well established, the percentage of lead in radioactive minerals should be a far more accurate method of deducing the age of the mineral than the calculation based on the volume of helium, for the lead formed in a compact mineral has no possibility of escape.

While the above considerations are of necessity somewhat conjectural in the present state of our knowledge, they are of value as indicating the possible methods of attacking the question as to the final products of the decomposition of the radioactive minerals. From a study of the data of analyses of radioactive minerals, Boltwood has suggested that argon, hydrogen, bismuth, and some of the rare earths possibly owe their origin to the transformation of the primary radioactive substances.

It does not appear likely that we shall be able for many years to prove or disprove experimentally that lead is the final product of radium. In the first place, it is difficult for the experimenter to obtain sufficient radium for working material, and, in the second place, the presence of the slowly transformed product radium D makes a long interval necessary before lead will appear in appreciable quantity in the radium. A more suitable substance with which to attack the question would be radium F (radiotellurium) or radiolead (radium D).

Constitution of the Radioelements

The view that the α particle is a helium atom suggests that the atoms of uranium and radium are built up in part of atoms of helium. If the final product of radium is lead, the radium atom could thus be represented by the equation, $Ra = Pb \cdot He_5$, while $Ur = Pb \cdot He_8$.

It must be borne in mind, however, that these compounds of helium are very different from ordinary molecular compounds. Both radium and uranium behave as elementary substances, which cannot be broken up by the application of physical or chemical forces at our command. These substances spontaneously break up at a rate that is independent of known agencies, and the disintegration is accompanied by the expulsion of a helium atom with enormous velocity. The energy liberated in the form of the kinetic energy of the expelled helium atoms is of quite a different order from that observed in molecular reactions, being at least one million times as great as that released in the most violent chemical combinations. It seems probable that the helium atoms are in very rapid motion within the atoms of uranium and radium, and for some reason escape from the atoms with the velocities which they possessed in their orbits. The forces that hold the helium atoms in place in the atom of the radioelements are so strong that no means at our disposal are able to effect their separation.

It seems probable that the α particles from thorium and actinium are also helium atoms, so that these substances must also be considered as compounds of some unknown substances with helium. Five *a* ray products are known to exist in thorium, and this would make the atomic weight of the residue of the thorium atom $232.5 - 5 \times 4$, or 212.5. The nearest known atomic weight to this is that of bismuth, viz. 208, and if thorium should lose six *a* particles instead of five, the atomic weight of the residue would be very nearly that of bismuth. This substance also fulfils the condition required for a transformation product of radioactive substances, for it is found in radioactive minerals, although only in small amount compared with that of lead in the old uranium minerals, where little thorium is present.

It thus appears that helium plays a most important part in the constitution of the radioelements, and it is not unlikely that helium as well as hydrogen may ultimately prove to be one of the more elementary units of which the heavy atoms are built up. In this connection it may prove more than a coincidence that a number of the atomic weights of the elements differ by nearly four units or by multiples of four units.

Many of the primary radioactive minerals were undoubtedly deposited at the surface of the earth 100 to 1000 million years ago, and since that time have been undergoing slow transformation. There is no evidence at hand that this process of degradation of matter is reversible under ordinary conditions at the surface of the earth. It seems, however, reasonable to suppose that under some conditions, existing possibly early in the earth's history, the converse process took place, and that the heavy atoms were built up from the lighter and more elementary substances.

It may happen that the conditions for the formation of heavy atoms may be found at the high pressures and temperatures existing deep in the earth. It has been suggested to me by Dr. Barrel, of Yale University, that the gradual building up of the heavy and more complex atoms of matter may be slowly taking place in the interior of the earth, and that this might possibly account for the undoubtedly high density of the matter in the interior of the earth, and also for the gradual shrinking of the earth as a whole.

While such suggestions are at present highly speculative, it appears not unreasonable to suppose that the formation of the radioactive matter may still be in progress deep in the earth, and that the radioactive deposits found at the surface to-day have been forced up from below in past ages.

CHAPTER IX

RADIOACTIVITY OF THE EARTH AND ATMOSPHERE

WE shall in this chapter briefly discuss the present state of our knowledge of the radioactive condition of the earth and atmosphere, and the possible bearing of the facts so far obtained on the problems of the electrical state of the atmosphere and on the internal heat of the earth.

ATMOSPHERIC RADIOACTIVITY

The remarkable development during the last few years of our knowledge of the radioactive and electrical state of the atmosphere is one of unusual interest, and although the interval for investigation has been short, a great deal of new and important information has been accumulated.

Nearly a century ago, Coulomb and others drew attention to the fact that a charged conductor placed inside a closed vessel lost its charge more rapidly than could be explained by the conduction of electricity along the insulating support. This was thought by Coulomb to be due to the molecules of air receiving a charge from contact with the charged rod and then being repelled from it. As early as 1850, Matteucci observed that the rate of loss of charge was independent of the potential of the charged body. By using insulators of quartz rods of different lengths and cross section, Boys in 1889 came to the conclusion that the loss of charge could not be explained by the imperfect insulation of the supports.

Shortly after science had become familiar with the ionization of gases by X-rays and uranium rays, the question of the cause of this loss of charge was independently attacked by Geitel[1] and C. T. R. Wilson,[2] using specially designed electroscopes to

[1] Geitel: Physik. Zeit., ii, p. 116 (1900).

[2] Wilson: Proc. Camb. Phil. Soc.: xi, p. 32 (1900); Proc. Roy. Soc., lxviii, p. 151 (1901).

measure the rate of discharge of a charged body inside a closed vessel. Both came to the conclusion that the gradual loss of charge was mainly due to an ionization of the air inside the closed vessel. Above a certain voltage, the rate of discharge was independent of the voltage, a result to be expected if the ionization was very weak. It was at first thought that this ionization in the gas was spontaneous and a property of the gas itself, but later work has modified this conclusion. It is now certain that a large part of the ionization observed in a clean metal vessel results mainly from the emission of ionizing radiations from its walls. A part is due to a very penetrating radiation of the γ ray type which is everywhere present on the surface of the earth. The amount of ionization of a gas inside a closed vessel depends on the nature and pressure of the gas and of the material of the vessel. In most cases the ionization falls off nearly proportionally with the pressure, and is approximately proportional to the density of the gas. Both of these results are to be expected if the ionization observed is due to radiations from the walls or to a penetrating type of radiation passing from the outside through the material of the vessel.

It must be borne in mind that the natural ionization observed in closed vessels is extraordinarily minute, and special precautions are usually necessary to measure it with accuracy. Assuming that the ionization in a small silvered glass vessel was uniform throughout its volume, C. T. R. Wilson found that not more than 30 ions were produced per second per cubic centimetre of the enclosed air. In a vessel of one litre capacity, the number of ions produced per second would be 30,000, or only about one third of the total number of ions produced in air by a single α particle emitted from radium. The expulsion of a single α particle per second from the walls of the vessel would thus more than account for the ionization observed.

After examining the discharge of electricity produced by air in closed vessels, Elster and Geitel turned their attention to the external air. They found that a charged body freely exposed to the open air lost its charge far more rapidly than when placed in a small closed vessel. Both positive and negative electricity is discharged, but generally at unequal rates, a positively charged body losing its charge somewhat more slowly than a negatively charged one. The ionization of the open air was examined by means of a portable electroscope. An insulated wire gauze was connected to the charged electroscope, and the rate of loss of charge of the electroscope was taken as a comparative measure of the number of ions in the air.

In the course of their experiments on closed vessels, Elster and Geitel noted that the rate of discharge increased for several hours after the introduction of fresh air. Such a result was known to occur when the radium or thorium emanation was mixed with the air. This led them to try a bold experiment to see if it were possible to extract a radioactive substance from the atmosphere. The writer had shown that a negatively charged wire exposed in the presence of the thorium emanation became strongly active. This experiment suggested the method of attacking the question.[1] A long wire was suspended on insulating supports outside the laboratory and charged negatively to a high potential by means of a static machine. After some hours the wire was removed and coiled round the top of an electroscope. There was an undoubted increase in its rate of discharge, showing that the wire possessed the new property of ionizing the gas. The effect died away after a time, and was small after a few hours' interval.

Further experiments showed that the wire had been made temporarily radioactive by exposure to the open air. The amount of activity observed was independent of the nature of the material of the wire, and in this respect the activity behaved quite similarly to the excited activity imparted to bodies in the presence of the radium and thorium emanations.

The active matter could be dissolved from the wire by rubbing it with leather soaked in ammonia. In this way an active substance was obtained capable of affecting a photographic plate through .1 mm. of aluminium and of producing weak phosphorescence on a screen of platinocyanide of barium.

[1] Elster and Geitel: Physik. Zeit., iii, p. 76 (1901).

Rutherford and Allan[1] showed that similar activity could be obtained from the open air in Montreal. The radiations consisted of α and β rays, the former being responsible for most of the ionization observed with bare wires. The activity of a wire made active by exposure to the atmosphere decayed at about the same rate as that of a wire made active by exposure to the radium emanation.

Bumstead and Wheeler[2] examined the radioactive state of the air at New Haven, and from a comparison of the rate of decay of the active wire with that of a wire made active by exposure to the radium emanation, showed conclusively that the activity observed in the air in that locality was mainly due to the radium emanation. A wire made active in the open air showed the initial rapid drop of activity due to radium A, and the curve of decay was identical with that due to the excited activity of radium. An emanation was obtained by boiling the soil and surface water at New Haven, which decayed at the same rate as the radium emanation.

By exposing wires for several days in the open air, Bumstead[3] also observed that, after the excited activity due to the radium emanation had disappeared, a part of the activity decayed much more slowly. This residual activity decayed at the same rate as the excited activity from thorium, showing conclusively that the thorium as well as the radium emanation was present in the air. Dadourian[4] showed that the soil at New Haven was impregnated with the thorium emanation. A hole was dug in the ground and the top closed. A negatively charged wire was exposed in the hole, and on removal it was found to show activity, which disappeared at the characteristic rate of the excited activity of thorium.

Such results show that the soil at New Haven must contain quite appreciable quantities both of thorium and radium. Since the thorium emanation has a very short life, it is only able to diffuse into the open air from a small depth of soil. The radium

[1] Rutherford and Allan: Phil. Mag., Dec., 1902.
[2] Bumstead and Wheeler: Amer. Journ. Science, Feb., 1904.
[3] Bumstead: Amer. Journ. Sci., July, 1904.
[4] Dadourian: Amer. Journ. Sci., Jan., 1905.

emanation, which has a much longer life, is able to emerge from a much greater depth.

In the meantime, C. T. R. Wilson[1] had found that rain was radioactive. Rain water was collected after a shower, and rapidly evaporated to dryness in a platinum dish, which was then placed under an electroscope. The activity was found to decay to half value in about 30 minutes.

Wilson in England, S. J. Allan and McLennan in Canada, independently showed that freshly fallen snow possessed a like property. The activity of snow, like that of rain, falls to half value in 30 minutes. This rate of decay is nearly the same as that observed for the excited activity of radium, several hours after removal of the emanation. Such a result suggests that the carriers of radium B and radium C become attached, probably by diffusion, to the water drops or snowflakes in their passage through the air. On evaporation, the active matter remains behind. A heavy fall of rain or a snowstorm must thus act as a means of temporarily removing a proportion of the radium B and C always present in the air.

Elster and Geitel found that the air in confined spaces, such as caves and cellars, was abnormally radioactive, and showed strong ionization. To show that these effects did not result from stagnant air alone, Elster and Geitel confined a large volume of air in an old steam boiler, but did not observe any increase of the ionization with time. Other experiments showed that the increased radioactivity in confined spaces, in contact with the earth, was due to the gradual storing of the radium emanation which diffused through the soil. In order to throw light on this question, Elster and Geitel[2] placed a pipe several feet deep in the earth, and by means of a pump sucked up some of the air imprisoned in the capillaries of the soil. This was found to be strongly active, and its activity decayed at about the same rate as that of air mixed with the radium emanation.

Similar results were observed by Ebert and Ewers[3] for the

[1] Wilson: Proc. Camb. Phil. Soc., xi, p. 428 (1902); xii, p. 17 (1903).

[2] Elster and Geitel: Physik. Zeit., iii, p. 574 (1902).

[3] Ebert and Ewers: Physik. Zeit., iv, p. 162 (1902).

air removed from the soil at Munich. Such results show conclusively that small quantities of radium are everywhere distributed throughout the surface soil of the earth. J. J. Thomson, Adams, and others examined the water obtained from deep wells and springs in England, and found that in some cases the water contained considerable quantities of the radium emanation, and in a few cases a trace of radium itself.

In the last few years, a very large amount of work has been done in examining the waters and sediments of mineral and hot springs for the presence of radioactive matter. H. S. Allan and Lord Blythswood found that the hot springs at Bath and Buxton contained appreciable quantities of a radioactive emanation. This was confirmed by Strutt, who found that not only was the radium emanation contained in the issuing water, but that the mud deposited by the springs contained traces of radium. It is of interest to note that helium has been observed amongst the gases evolved by these springs, and it would appear probable that the waters in their passage to the earth pass through a deposit of radioactive minerals.

Himstedt found the radium emanation in the thermal springs at Baden Baden, while Elster and Geitel found also small traces of radium in the mud deposited by them. A large number of springs have been examined by different observers in England, Germany, France, Italy, and the United States, and in nearly all cases the radium emanation has been found in the water, often in easily measurable amount. Elster and Geitel found that the mud or "fango" deposited from the hot springs at Battaglia, Italy, was abnormally radioactive, and a close examination showed that the activity was due to radium. They calculated that, weight for weight, it contains almost one thousandth of the radium to be obtained from the Joachimsthal pitchblende.

While the activity of most of the waters of hot springs is due in most cases to the presence of radium or its emanation, Blanc[1] has observed one notable exception in which the activity is mainly due to thorium. The sediment of the waters at Salins-Moutiers

[1] Blanc: Phil. Mag., Jan., 1905.

was abnormally active, and was found to give off considerable quantities of the thorium emanation. Blanc, however, was unable to detect analytically the presence of thorium, although, from the amount of emanation given off, a considerable quantity should have been present. It seems not unlikely that the activity observed was due not to the primary substance thorium, but to its product, radiothorium, which was discovered by Hahn (see page 68). This would give rise to thorium X and the thorium emanation, but would be present in too minute a quantity to be determined chemically.

Elster and Geitel observed that natural carbonic acid obtained from great depths of old volcanic soil contained the radium emanation, while McLennan and Burton found considerable quantities of radium emanation in the petroleum from a deep well in Ontario, Canada.

In most cases where spring water comes from great depths, and especially if the water is hot, radioactive matter is found to be present in abnormal amount compared with that found in the soil itself. Such a result is not unexpected, for water, and particularly hot water, would tend to dissolve traces of radioactive matter in the strata through which it passes, and also to become impregnated with the radium emanation. In special cases, it may happen that the water has passed through a deposit of radioactive minerals, and in such a case, a very strong activity is to be expected.

Elster and Geitel have made an extensive examination of different soils for radioactivity, and have found traces of radioactive matter in nearly every case. The activity is most marked in clayey soils, and is apparently due in many cases to the presence of small traces of radium. The observations as a whole show clearly that radioactive matter is extraordinarily diffused in nature, and it is difficult to find any substance that does not contain a minute trace of radium. It does not appear likely that uranium or radium differ in this respect from the inactive elements.

The presence of radium can be noted by the electric test, where chemical analyses would fail to detect the presence of rare inactive elements, although they may exist in considerably greater quantity than the radium itself. On general grounds, such a wide diffusion of radioactive matter is not surprising, for the soil of the earth at any point should contain a fairly thorough admixture of a great majority of the elements found in the earth, the rare elements being present in only minute proportion.

There can be no reasonable doubt that the radioactive matter observed in the atmosphere is mainly due to the emanation of radium and its transformation products, and probably in some localities to traces of the emanation of thorium and actinium. The supply of radioactive matter in the atmosphere is kept up mainly by the diffusion and transpiration of the emanations through the soil, while no doubt a part is supplied by the action of springs and by the release of imprisoned gases.

On account of its comparatively slow rate of change, it is to be expected that the amount of the emanation of radium will predominate over the other emanations in the atmosphere, for the short life of the emanations of actinium and thorium prevent their reaching the surface from any appreciable depth. While it is probable that the supply of emanation from the earth to the atmosphere varies in different localities, the action of winds and air currents generally should tend to distribute the emanation from point to point and to make its distribution more uniform.

All observers have noted that the amount of excited activity to be obtained under definite conditions from the atmosphere is very variable, and often alters considerably during a single day. Elster and Geitel made a detailed examination of the effect of meteorological conditions on the amount of active matter in the atmosphere. The experiments were made at Wolfenbüttel, Germany, and were continued for twelve months. On an average, the amount of active matter increased with a lowering of the temperature. Below 0° C., the average was 1.44 times as great as above 0° C. A falling barometer increases the amount of active matter. The effect of a change of pressure is intelligible when it is remembered that a lowering of the pressure tends to cause the emanation in the capillaries of the soil to be drawn to the surface.

If the emanation observed in the atmosphere is entirely drawn from the soil, the amount in the air over the sea in mid-ocean should be much smaller than on land, for the water will not allow the emanation to escape from the earth's crust into the atmosphere. Observations so far made indicate that the amount of active matter in the air falls off near the sea. For example, the amount on the Baltic coast was found by Elster and Geitel to be only about one third of that found inland, but no systematic examination has yet been made of the amount of active matter in the atmosphere at great distances from land.

Amount of the Radium Emanation in the Atmosphere

Most of the experiments on the amount of active matter in the air have been qualitative in character, but it is obviously important to obtain some idea of the amount of the radium emanation present in the atmosphere. Since the amount in the atmosphere is kept up by a constant supply of fresh emanation from the earth, it is convenient to express the amount of radioactive matter in the atmosphere in terms of the amount of freely emanating radium bromide which is required to keep up the supply.

Some interesting experiments in this direction have recently been made by A. S. Eve[1] at Montreal. The radioactive state of the air in the neighborhood of Montreal appears to be normal, and the number of ions present per cubic centimetre of the outside air is about the same as that observed at different localities in Europe.

Some experiments were first made in a large iron tank in the Engineering Building of McGill University. This tank was 8.08 metres high and 1.52 metres square, with a total volume of 18.7 cubic metres. In order to determine the amount of excited activity to be obtained from the tank, a long insulated wire was suspended in its centre, and kept at a constant potential of –10,000 volts for three hours. The wire was then rapidly

[1] A. S. Eve: Phil. Mag., July, 1905.

removed, and coiled round a frame attached to an electroscope. The rate of fall of the gold leaves then served as a measure of the amount of active matter deposited on the wire.

A similar experiment was then made in a small zinc cylinder of volume 76 litres. The emanation derived from 2×10^{-4} milligrams of radium bromide was introduced into the cylinder and mixed with the air. The excited activity was concentrated, as before, on a negatively charged wire, and after removal was measured with the same apparatus employed in the first experiment. Knowing the rate of discharge of the electroscope for the active deposit obtained from a known quantity of radium, a comparison of the results obtained with the large tank at once gave the amount of emanation present in it. In this way, it was calculated that one cubic kilometre of the air, containing the same amount of emanation per unit volume as that observed in the large tank, was equivalent to the emanation supplied by .49 grams of pure radium bromide.

The tank in these experiments was in free communication with the open air, and the amount of the excited activity was unaltered when the air from the room was forced through the tank. It thus seems reasonable to suppose that the air within the tank contained the same amount of emanation per unit volume as the outside air. No radioactive matter had ever been introduced into the building where the tank was used, and, as we shall see later, the rate of production of ions per cubic centimetre of the tank was lower than that ever previously recorded.

In order to verify this point, however, experiments were made on another large zinc cylinder placed on the College Campus, with its ends in free communication with the air. The active deposit was collected on a wire suspended along the axis of the tank, and tested as before. The average amount, however, was found to be only from one third to one fourth of that observed for an equal volume of the large tank. No adequate cause could be assigned for this discrepancy between the results for the experiments in the iron tank and in the zinc cylinder, unless the charged wire is unable for some reason to collect all the active deposit from the cylinder placed in the open air.

On certain assumptions we can form a rough estimate of the amount of radium emanation existing in the atmosphere. Suppose the emanation is uniformly distributed in a spherical layer 10 kilometres deep around the earth, and that the emanation per cubic kilometre is uniform, and equal to that observed at Montreal. The surface of the earth is about 5×10^8 square kilometres, and the volume of the shell 10 kilometres thick is 5×10^9 cubic kilometres. Taking the estimated value, .49 grams per cubic kilometre, found from experiments on the large tank, the amount of emanation in the atmosphere corresponds to 2.5×10^9 grams or 2,460 tons of radium bromide.

Now, about three quarters of the earth is covered with water, through which no emanation can escape to the surface. If the emanation arises from the land alone, the amount is thus reduced to about one quarter of this, or 610 tons. Taking the value obtained from measurements in the cylinder in the outside air, the amount is found to be about 170 tons.

Several observers have shown that the amount of excited activity present in the air at high altitudes is equal to, if not greater than, the amount observed on the plains. It thus appears probable that in supposing that the emanation is distributed to an average height of 10 kilometres in the atmosphere we are well within the truth. Until a very complete radioactive survey of the atmosphere has been made, such calculations are of necessity somewhat uncertain, but they certainly serve to give the right order of magnitude of the quantities involved.

Since the emanation is half transformed in about four days, it cannot diffuse into the air from any great depth in the earth, so that the main supply of the emanation must come from a superficial layer of the earth not many metres in thickness. A part of the supply is probably due to deep seated springs, which may bring up the emanation from greater depths, but the amount so supplied is probably small compared with that escaping directly through the pores of the soil.

We thus arrive at the important conclusion that a very considerable quantity of radium, measured by hundreds of tons, is distributed over the earth within a few metres of its surface.

It is for the most part, however, distributed in such infinitesimal quantities that its presence can be detected only by the aid of the electric method.

Eve (*loc. cit.*) found that a wire about one millimetre in diameter charged to –10,000 volts, and suspended about 20 feet above the ground, was only able to collect the active deposit from the air in a cylindrical volume of radius lying between 40 and 80 cms. This collecting distance is small compared with that to be expected for such a high voltage, for the writer has shown that the positively charged carriers of the active deposit of radium and of thorium travel in an electric field at about the same velocity as the ion, *i. e.*, they move with a velocity of about 1.4 cms. per second under a potential gradient of one volt per cm. It seems probable that the carriers of the active deposit, which must remain suspended a long time in the atmosphere, adhere to the comparatively large dust nuclei always present in the air, and consequently move very slowly in an electric field, so that the carriers can only be drawn in from the immediate vicinity of the charged wire.

The Penetrating Radiation at the Earth's Surface

Since the radium emanation is everywhere present in the surface of the earth and in the atmosphere, its transformation product radium C must give rise to γ rays, and these rays must come in all directions from the earth and atmosphere. The presence of such a penetrating radiation at the earth's surface was independently observed by McLennan [1] and H. L. Cooke,[2] in Canada. McLennan worked with a large vessel, and observed that the ionization of the air inside diminished about 37 per cent when the vessel was surrounded by a thickness of 25 cms. of water. Cooke worked with a small brass electroscope of about one litre capacity. The rate of discharge of the electroscope fell about 30 per cent when completely surrounded by a lead screen 5 cms. thick. No further diminution was observed by

[1] McLennan: Phys. Rev., No. 4, 1903.

[2] Cooke: Phil. Mag., Oct., 1903.

placing a ton of lead around the apparatus. The radiation is of about the same penetrating power as the γ rays from radium, and can be observed in the open air as well as in a building. By placing blocks of lead in different positions in regard to the electroscope, it was found that the radiation came about equally from all directions, and was the same during the day as at night. Such results are to be expected if the penetrating rays come equally from the radioactive matter distributed in the earth and atmosphere. The magnitude of the ionizing effect due to the penetrating rays is, however, much greater than that due to the γ rays from the amount of radium emanation in the atmosphere calculated by Eve. It seems not unlikely that these penetrating rays may be given out by matter in general as well as by radioactive bodies.

Electrical State of the Atmosphere

It has long been known, from observation of the potential gradient in the atmosphere, that the upper layers of the atmosphere are generally positively charged in regard to the earth. There is consequently an electric field always acting between the earth and the upper atmosphere. Since there is a distribution of ions in the lower regions of the atmosphere, there must consequently be a steady movement of negative ions upwards and of positive ions downwards towards the earth. Since the carriers of the active deposit of radium have a positive charge, they must tend to be deposited on the surface of the earth. Each blade of grass and each leaf must consequently be coated with an invisible deposit of radioactive material. A hill or mountain top tends to concentrate the earth's field at that point, and there should be a greater amount of active matter deposited on its surface than on an equal area on the plains. This is in agreement with the observations of Elster and Geitel, who found that the ionization of the air on a mountain top was greater than on a lower level.

A large number of observations have been made of the relative number of ions in the air at various localities under different meteorological conditions. Many experimenters have used the "dissipation apparatus" constructed by Elster and Geitel. This consists of an open wire gauze connected with an electroscope. The rate of discharge of the electroscope is separately observed when charged positively and negatively. While this apparatus has proved of value in preliminary work on the ionization of the atmosphere, the results obtained are only comparative, and do not readily lend themselves to quantitative calculations. The effect of wind is very marked in such an apparatus, and the rate of dissipation is always higher when a wind is blowing.

A very useful portable instrument for determining the actual number of positive and negative ions per cubic centimetre of the air has been devised by Ebert.[1] By means of a fan driven by clockwork, a steady current of air is drawn between two concentric cylinders. The inner cylinder is insulated and connected with a direct reading electroscope. The length of the cylinder is so adjusted that all the ions in the air are drawn to the electrodes in their passage through the cylinder. Knowing the capacity of the instrument, the velocity of the current of air, and the constants of the electroscope, the number of ions per c.c. of the air can easily be deduced. When the inner cylinder is charged positively, the rate of discharge of the electroscope is a measure of the number of negative ions in the air, and *vice versa*.

Measurements by Eberts and others show that the actual number of ions per c.c. of air is subject to considerable fluctuations and is dependent on meteorological conditions. The number usually varies between five hundred and several thousands, and the number of positive ions is nearly always greater than the number of negative.

Schuster[2] observed that the number of ions per c.c. in the air in Manchester varied between 2300 and 3700. These numbers give the equilibrium number of ions in the air when the rate of production of fresh ions is balanced by the rate of their recombination. If n_1, n_2 are the number of positive and negative ions respectively per c.c. of air, and q the rate of production per c.c.

[1] Ebert: Physik. Zeit., ii, p. 662 (1901); Zeitschr. f. Luftschiff-fahrt, iv, Oct., (1902).

[2] Schuster: Proc. Manchester Phil. Soc., p. 488, No. 12, 1904.

per second, then $q = a\, n_1\, n_2$, where a is the coefficient of recombination of the ions. By a slight modification of the apparatus of Ebert, Schuster was able to determine the value of a for the air under the normal conditions of experiment, and deduced that the value of q in Manchester varied between 12 and 39.

The apparatus of Ebert was designed to measure the number of free ions in the air which have the same mobility as the ions produced by X-rays or the radiations from active bodies. The velocity of the ions produced in air have been directly measured by Mache and von Schweidler. The positive ion moves 1.02 cms. per second, and the negative 1.25 cm. per second, for a potential gradient of one volt per cm. These velocities are slightly slower than those observed for the ions produced in dust-free air by X-rays or the rays from radioactive substances.

In addition to these swiftly moving ions, Langevin[1] has shown that a number of slowly moving ones are also present, which travel too slowly in an electric field to be removed by the electric field used in the apparatus of Ebert. These ions move with about the same velocity as the ions observed in flame gases some distance from the flame. By using much stronger fields, Langevin has determined the number of these heavy ions present in the air, and concludes that they are about forty times as numerous as the swiftly moving ones. It is possible that these slowly moving ions are formed by the deposition of water round the ion to form a minute globule, or by the adherence of the ion to the dust which is always present in the air.

Since there is undoubtedly a continuous production of ions in the air near the earth, it is a matter of great importance to determine the cause or causes of this ionization. The most obvious cause is the presence of radioactive matter in the atmosphere. But is the amount present capable of producing the ionization observed? In order to throw light on this important point, Eve (*loc. cit.*) made the following experiment. The large iron tank, previously described on page 204, was used. An insulated cylindrical electrode passed down the centre of the

[1] Langevin: Comptes rendus, cxl, p. 232 (1905).

tank and was connected to an electroscope. The electrode was charged to a sufficient potential to obtain the saturation current, which is a measure of the total number of ions produced per second. A wire charged to –10,000 volts was then suspended in the tank and the active deposit collected from it for a definite time. The activity imparted to the wire was measured immediately after removal with an electroscope.

An exactly similar set of experiments was then made with a much smaller zinc cylinder, the air of which was artificially supplied with the radium emanation obtained by blowing air through a radium solution. The saturation current was measured, and also the amount of the activity imparted to a central electrode under the same conditions as in the large tank. If the ionization in the large tank is due entirely to the presence of the radium emanation in it, then the ratio of the saturation currents in the two tanks should be equal to the ratio of the activities imparted to the collecting wires under the same experimental conditions. This must obviously be the case, since the saturation current serves as a measure of the amount of emanation present, and so also does the activity imparted to the collecting wire.

The ratio of the activity on the collecting wire in the iron tank to that in the emanation cylinder was found to be about 14 per cent less than the ratio of the corresponding saturation currents. Considering the difficulty of such experiments, the agreement is as close as could be expected, and indicates that the greater part, if not all, of the ionization observed in the iron tank was due to the presence of the radium emanation.

Since there was every reason to believe that the air in the tank contained the same amount of emanation as the outside air, this result indicates that the production of ions in the outside air is mainly due to the radioactive matter contained in it. Before such a conclusion can be considered as established, experiments of a similar character must be made in various localities. We are, in any case, justified in assuming that the radioactive matter in the air plays a very important part in the production of ions observed in the atmosphere near the surface of the earth.

It is of interest to record that Eve found the number of ions produced per c.c. per second in the iron tank to be 9.8. This is the smallest rate of production of ions yet recorded for a closed vessel. Cooke observed a value as low as 20 for a well cleaned brass electroscope of about one litre capacity.

If the radioactive matter in the air is the cause of its ionization, there should be a constant proportion between the rate of production of ions in the air and the excited activity on the collecting wire. The data so far collected by various observers appear to contradict such a connection. It is doubtful, however, whether the measurements actually supply the data required.

There seems to be no doubt that the recombination constant of the ions depends greatly on meteorological conditions, and on the freedom of the air from nuclei. The variation of this constant affects the equilibrium number of ions in the air determined by the apparatus of Ebert. In a similar way, the excited activity imparted to a charged wire in the open air will probably depend upon atmospheric conditions, although the amount of emanation present may not have been changed. Before any definite conclusion can be reached, it will be necessary to take all these factors into account. A large number of observations have been made in Germany on the effect of meteorological conditions on the amount of dissipation measured by Elster and Geitel's apparatus. We have already mentioned the effect of a rising or falling barometer, in producing well marked variations in the amount of active matter in the air. The relation between potential gradient and dissipation has been studied by Gockel and Zölss. The latter finds that the potential gradient varies in a marked manner with the dissipation. A high potential gradient is accompanied by a low value of the dissipation, and *vice versa*. A similar relation between the potential gradient and the amount of ionization determined by Ebert's apparatus has been observed by Simpson in Norway. Elster and Geitel, and Zölss have shown that the dissipation increases with the temperature. Simpson found that at Karasjoh in Norway, the average between temperatures of 10° C. and 15° C. was about six times as great as that between –40° C. and –20° C.

A very complete series of observations on the annual variation of the potential gradient, ionization, and dissipation, was made by Simpson[1] at Karasjoh, in Norway, situated within the Arctic circle, in latitude 69°. These results are of special interest, for between November 26 and January 18 the sun did not rise above the horizon, while between May 20 and July 22 the sun did not fall below the horizon.

The absence of the sun's rays apparently had no marked effect on the magnitude of the quantities measured. There was on an average a steady rise of the potential gradient between October and February, and a steady fall of the ionization during the same period. Such results indicate that the sun's rays have little if any direct effect on the ionization of the air.

It is impossible here to discuss the numerous speculations that have been advanced to account for the presence of a strong positive charge in the upper atmosphere. This positive charge must be steadily supplied from some source, for otherwise it would be rapidly discharged by the ionization currents between the upper and lower atmosphere. Our knowledge of the electrical state of the upper atmosphere is at present too imperfect to enable us to determine whether this distribution of the charge is due to an effect of radiations from the sun, as some have supposed, or to a separation of the positive and negative ions continuously produced in the atmosphere.

Internal Heat of the Earth

The problem of the origin of the earth's internal heat has been a subject of intermittent discussion for more than a century. The most plausible and the generally accepted view is that the earth was originally a very hot body, and in the course of millions of years has cooled down to the present state. This process of cooling is supposed to be still continuing, with the result that the earth will ultimately lose its internal heat by radiation into space.

On this theory, Lord Kelvin bases his well known deduction of the age of the earth as a habitable planet. From observations

[1] Simpson: Trans. Roy. Soc. Lond. A, p. 61, 1905.

of bores and mines, it has been found that the temperature of the earth increases steadily from the surface downwards, and on an average this temperature gradient is found to be about 1/50° F. per foot, or .00037° C. per cm. In order to obtain an estimate of the maximum age of the earth on this theory, Kelvin supposed that the earth was initially a molten mass. By an application of Fourier's equation, it is possible to deduce the temperature gradient at the surface of the earth at any time after the cooling began, provided the initial temperature and the average conductivity for heat of the materials of the earth are known. Taking the most probable value of these numbers, Kelvin in his original calculations found that the time required for the earth to cool from the temperature of a molten mass of rock to its present state was about 100 million years. In later calculations, using improved data, this estimate has been cut down to about 40 million years.

On this theory, life cannot have existed on the earth for more than 40 million years. This period has been thought by many geologists and biologists to be far too short to account for the processes of organic and inorganic evolution, and for the geologic changes observed in the earth, and such a serious curtailment of the time at their disposal has given rise to much controversy. On the theory on which Kelvin bases this calculation, there can be little doubt of the probable correctness of this estimate of the age of the earth, although the experimental data on which the calculations were based are of necessity somewhat imperfect. This theory, however, assumes that the earth is a simple cooling body, and that there has been no generation of heat from internal sources, for Lord Kelvin pointed out that the possible heat developed by the earth's contraction or by ordinary chemical combination is not sufficient to affect appreciably the general argument.

The discovery of the radioactive bodies, which emit during their transformation an amount of heat at least one million times greater than that observed in ordinary chemical changes, throws quite another light on this question. We have seen that radioactive matter is everywhere distributed through the surface of the earth and in the atmosphere, and that the amount of radium existing close to the surface is of the order of several hundred tons.

It is of interest to calculate how much radium must be uniformly distributed in the earth in order to compensate for the present loss of heat from the earth by conduction to the surface. The heat in gram calories per second lost by conduction to the surface of the earth is given by

$$Q = 4 \pi R^2 K T,$$

where R = radius of the earth, K the heat conductivity of the earth in C. G. S. units, and T the temperature gradient. Let X be the average amount of heat liberated per second per cubic centimetre of the earth's volume, owing to the presence of radioactive matter. If the heat, Q, supplied per second is equal to that lost by conduction to the surface, then

$$X \tfrac{4}{3} \pi R^3 = 4 \pi R^2 K T,$$

$$\text{or } X = 3 \frac{K T}{R}.$$

Taking the average value of $K = .004$, the value taken by Lord Kelvin, and $T = .00037$, then

$$X = 7 \times 10^{-15} \text{ gram calories per second}$$
$$= 2.2 \times 10^{-7} \text{ gram calories per year.}$$

Now one gram of radium in radioactive equilibrium emits 876,000 gram calories of heat per year. Consequently the presence of radium to the amount of 2.6×10^{-13} grams per c.c., or 4.6×10^{-14} grams per unit mass would compensate for the heat lost by conduction.

In this calculation, the amount of radioactive matter present has been expressed in terms of radium. There is no doubt that uranium, thorium, and actinium are also present, but the heating effects of these are expressed in terms of radium. On this view, the total heating effect of radioactive matter present in the earth is equivalent to that of about 270 million tons of radium.

Such an estimate does not appear to be excessive when it is remembered that there is undoubted evidence that several hundred tons of radium are present in a thin shell at the earth's surface. Taking the estimate of Eve that about 600 tons of radium are required to keep up the supply of emanation in the atmosphere over the land, it can be calculated that this comes from a superficial layer of the earth, about 18 metres in depth. This is based on the assumption that the radium in the earth is distributed uniformly in the amount previously calculated. Such a thickness is of the order of magnitude to be expected from general considerations.

The experiments of Elster and Geitel have shown that radioactive matter is found in rocks and in soils in about the amount required by this theory. The heating effect of this radioactive matter must undoubtedly be taken into account in deductions based on the temperature gradient observed at the earth's surface. If the calculated amount of radium were distributed uniformly in the earth, the temperature gradient would remain constant as long as the supply of radioactive matter remains unchanged. If the radioactive matter existed near the surface of the earth in amounts greater than this mean value, the temperature gradient would be correspondingly greater than the observed value.

While the data on which these deductions are based is of necessity somewhat meagre, the evidence so far obtained is sufficiently strong to cast grave doubts on the validity of the calculations of the age of the earth, based on the view that it is a simple cooling body. The temperature gradient observed in the earth to-day may have remained sensibly constant for millions of years in consequence of the steady generation of heat in the earth.

It does not seem feasible on this theory of the maintenance of the earth's heat to fix with any certainty the age of the earth. The radium present in the earth is derived from the parent substance uranium, and on this theory uranium must exist in the earth in the proportion of about one part in fifty million. This proportion does not seem excessive from present data. The life of uranium is about 1000 million years, so that if the internal heat of the earth were due entirely to uranium and radium the temperature gradient 1000 million years ago would only be about twice that observed to-day.

It has already been pointed out that some of the uranium minerals are undoubtedly several hundred million years old, and the evidence suggests that some of them are of still greater antiquity. The evidence deduced from radioactive data alone points very strongly to the conclusion that on the lowest estimate, the earth is several hundred million years old.

The radioactive data do not of themselves enable us to decide whether the earth was originally a very hot body or not. The theory that the earth was originally a molten mass seems to have been largely the outcome of an attempt to explain the internal heat of the earth. Some geologists, notably Professor Chamberlin of Chicago, have long upheld the view that the geologic evidence by no means supports such a conclusion. It is not possible here, however, to do more than mention this interesting possibility.

Radioactivity of Ordinary Matter

It is a matter of general experience that every physical property discovered for one element has been found to be shared by others in varying degrees. For example, the property of magnetism is most marked in iron, nickel, and cobalt, but every substance examined has been found to be either feebly magnetic or diamagnetic. It might thus be expected on general principles that the property of radioactivity which is so marked in a substance like radium would be shown by other substances.

A preliminary examination at once showed that if ordinary matter was radioactive at all, it was only so in a minute degree, but later work by McLennan, Strutt, Campbell, Wood, and others, has shown that ordinary matter does possess the property of ionizing the gas to a small extent. Campbell[1] in particular has carefully examined this question, and the evidence obtained by him affords very strong proof that ordinary matter

[1] Campbell: Phil. Mag., April, 1905; Feb., 1906.

does possess the property of emitting ionizing radiations, and that each element emits radiations differing both in character and intensity.

Experiments on this subject are very difficult, as the ionization currents measured are extraordinarily minute. The effects are very complicated as each substance emits α rays, and penetrating rays, and the latter in some cases give rise to a marked secondary radiation.

Campbell concludes that the α rays emitted from lead have a range of ionization in air of about 12.5 cms., while those from aluminium have a range of only 6.5 cms. On an average the α rays emitted from ordinary matter have a considerably greater range in air than the α rays from radium. A sample of the lead employed was dissolved in nitric acid and tested by the emanation method for the presence of radium, but not the slightest trace was observed.

It does not necessarily follow that these α particles are identical in mass with the α particles of radium. They may possibly be hydrogen atoms, for if the α particles from ordinary matter were helium atoms we should expect, for example, to find helium in lead.

If the expulsion of α particles be taken as evidence of atomic disintegration, a simple calculation shows that the life of ordinary matter is of the order of at least one thousand times that of uranium, *i. e.* not less than 10^{12} years.

CHAPTER X

PROPERTIES OF THE α RAYS

In the previous chapters we have shown how prominent a role the α rays play in radioactive phenomena, as compared with the more penetrating β and γ rays. Not only are they responsible for most of the ionization observed in the neighborhood of radioactive matter, but they are also directly concerned with the rapid emission of heat energy from these substances; in addition, they generally accompany the transformation of the different types of radioactive matter, while the β and γ rays are emitted only in the case of a few products. Finally, we have seen that there is good reason to believe that the α particle is to be identified with an atom of helium.

In this chapter, we shall outline in some detail the more important properties possessed by the α rays, and especially by the α rays emitted by radium and its products. On account of their great intensity, the α rays from radium have been more easily studied than the corresponding rays from feebly radioactive substances like uranium and thorium. At the same time, the evidence so far obtained indicates that the α particles from all the radioactive substances have the same mass, and differ for each product only in their initial velocity of projection.

The α rays differ from the β and γ rays in the ease with which they are absorbed by matter and by the comparatively large ionization they produce in the air near to a radioactive body. By examining the effect of adding thin screens of metallic foil over radioactive matter, it was found that the α rays from the radioactive substances differed in penetrating power.

We shall see later that the α rays from radium are completely cut off by a layer of aluminium foil of thickness .04 mms., or by a layer of air of thickness 7 cms. The ionizing action of the α rays is consequently confined within a short distance, while that of the β rays extends for several metres, and that of the γ rays for several hundred metres.

The α rays were at first thought to be non-deflectable by a magnetic field, for the application of a magnetic field sufficiently strong to bend away completely the β rays had no appreciable effect on the α rays.

In 1901, the writer began experiments by the electric method, to see if the α rays could be deflected in a strong magnetic field, but with the weak preparations of radium (activity 1000) then available, the electric effects were too small to push the experiments to the necessary limit. Later, in 1902, using a preparation of activity 19,000, the experiments [1] were successful, and the α rays were found to be deflected in passing through both a magnetic and an electric field.

The direction of the deflection was opposite to that for the β rays, and this indicated that the α rays consisted of a flight of positively charged particles. By measurement of the amount of deflection of the rays in passing through magnetic and electric fields of known strengths, the mass and velocity of the α particle were determined. The value of e/m — the ratio of the charge carried by an α particle to its mass — was found to be about 6×10^3, while the maximum velocity v of the particle was found to be 2.5×10^9 cms. per second.

Since the ratio e/m for the hydrogen atom is about 10^4, this result indicated that the α particle was atomic in size, and, assuming that the α particle carried the same charge as a hydrogen atom, had a mass about twice that of the hydrogen atom. It must be remembered that the amount of the deviation of the rays in a given magnetic field is minute in comparison with that of the β rays. For example, the swiftest α particle projected from radium at right angles to a magnetic field of 10,000 C. G. S. units describes the arc of a circle of 40 cms. radius. The swiftest β particle from radium, which is projected with 96 per cent of the velocity of light, describes under similar conditions a circle of about 5 mms. radius.

[1] Rutherford: Physik. Zeit., iv, p. 235 (1902); Phil. Mag., Feb., 1903.

Becquerel[1] confirmed the magnetic deflection of the α rays from radium by means of the photographic method, and also showed that the α rays from polonium have a similar property. Using some pure radium bromide as a source of rays, Des Coudres[2] measured the deflection of a pencil of α rays in a vacuum after passing through a magnetic and electric field. He found e/m to be 6.3×10^3 and the velocity to be 1.64×10^9 cms. per second. The values of e/m obtained by Rutherford and Des Coudres were in good agreement, but the velocities varied considerably. In the experiments of Des Coudres the α rays were passed through a screen of aluminium. It will be seen later that this reduces the velocity of the α particles, and that the correct velocity of the swiftest α particle from radium is about 2×10^9 cms. per second, or about 1/15 of the velocity of light.

In 1905, the question was again attacked by Mackenzie,[3] using pure radium bromide as a source of rays. A photographic method was employed, in which the α rays fell on a glass plate coated on its lower surface with zinc sulphide. A photographic plate was placed on the upper side of the glass plate, and was acted on by the light from the scintillations produced by the α rays in the zinc sulphide screen immediately below it. The deflection of the pencil of rays was observed as before after passing through a magnetic and electric field. The α rays were found to be unequally deflected by a magnetic field, showing that the α particles varied either in mass or velocity. This dispersion of the rays by a magnetic and electric field made it difficult to deduce the constants of the rays with accuracy. Taking the mean value for the dispersion of the deflected pencils, he found the value of e/m to be 4.6×10^3, and the velocity of the α particles to vary between 1.3×10^9 and 1.96×10^9 cms. per second, on the assumption that the α particles all carry the same charge and have the same mass.

The importance of an accurate determination of e/m for the α particle had long been recognized because of the light it

[1] Becquerel: Comptes rendus, cxxxvi, pp. 199, 431 (1903).
[2] Des Coudres: Physik. Zeit., iv, p. 483 (1903).
[3] Mackenzie: Phil. Mag., Nov., 1905.

would throw on the question whether the α particle is an atom of helium. In all the methods so far described, a thick layer of radium in radioactive equilibrium was employed as a source of rays. On the theory of absorption of the α rays, put forward by Bragg and Kleeman, which will be discussed later, it was recognized that the α rays emitted from a thick or a thin layer of radium must consist of α particles moving at different speeds. The use of a complex pencil of rays was open to very serious objections, for it was impossible to know whether the rays most deflected in a magnetic field corresponded to the most deflected rays in an electric field or not.

The simplest method of accurately determining the value of e/m is to use a homogeneous source of rays, *i. e.*, to use a radioactive substance in which all the α particles escape at the same speed. The writer found that a wire made active by exposure to the radium emanation completely satisfied these essentials. The active matter, consisting of radium A, B, and C, is deposited in an extremely thin film on a negatively charged wire exposed in the presence of the emanation. After three hours' exposure, the activity of the wire reaches a maximum value. After removal, radium A, which has a three minute period, is rapidly transformed, and has practically disappeared after fifteen minutes. The activity remaining is then entirely due to radium C. The α particles from radium C are all expelled with identically the same velocity, for there is no appreciable dispersion of the rays in a magnetic field. The particles projected into the wire are completed absorbed, and those which escape do not suffer in velocity in their passage through the very thin film of intensely active matter.

Using 10 to 20 milligrams of radium in solution, a wire one centimetre long can be made extremely active by an arrangement similar to that shown in Fig. 24, page 100. The wire produces a strong photographic impression on a plate brought near it. The chief drawback to such a source of rays is that the intensity of the rays falls off rapidly, and two hours after removal is only 14 per cent of the initial value.

The apparatus shown in Fig. 42 is very convenient for the

determination of the magnetic deflection of the rays. An active wire is placed in a groove A. The rays pass through a narrow slit B and fall on a small piece of photographic plate at C. The apparatus is enclosed in a cylindrical vessel P, which can be rapidly exhausted of air. The apparatus is placed between the pole pieces of a large electromagnet, so that the magnetic field is parallel to the direction of the wire and slit, and uniform over the whole path of the rays. The electromagnet is excited by a constant current, which is reversed every ten minutes. On developing the plate two well defined bands are observed corresponding to the pencils of rays which have been deflected equally on opposite sides of the normal.

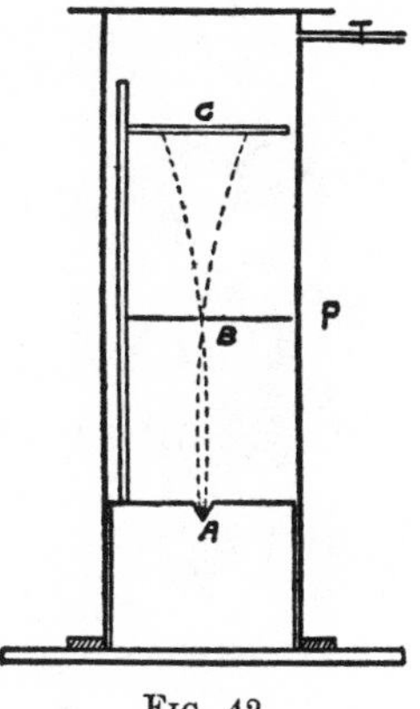

FIG. 42.

Apparatus for determining the amount of deflection of a pencil of α rays in a strong magnetic field.

If ρ is the radius of curvature of the circle described by the rays in a uniform field of strength H, then $H\rho = \frac{mv}{e}$, where v is the velocity of the rays, e the charge on the particle, and m its mass.

Let d = deflection of the rays from the normal measured on the photographic plate,

a = distance of plate from slit,
b = distance of slit from the source.

Then by a property of the circle, if the deflection d is small compared with a,

$$2\rho d = a(a+b).$$

Consequently,

$$\frac{mv}{e} = H\rho = \frac{Ha(a+b)}{2d}.$$

In the actual photographs, using the wire as a source of rays, the traces of the pencil of rays stand out clearly with well defined edges, so that the value $2d$, the distance of the inside edge of one band to the outside edge of the other, can easily be measured.

The value of $H\rho$ for the α particles emitted from radium C was found in this way to be 4.06×10^5. In a field of 10,000 C. G. S. units, the α particle consequently describes a circle of radius 40.6 cms.

Retardation of the Velocity of the α Particle in passing through Matter

It was found by the writer[1] that the velocity of the α particle diminishes in its passage through matter. This is most simply shown by a slight modification of the experimental arrangement, described above, which has been used by Becquerel. By means of mica plates, placed at right angles to the slit, the apparatus is divided into two equal parts. One half of the photographic plate is acted on by the rays from the bare wire, and the other by the rays which have passed through an absorbing screen placed over the wire.

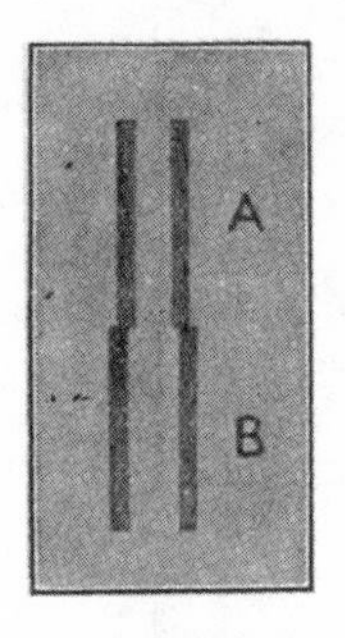

Fig. 43.

Retardation of the velocity of the α particles in passing through matter.

A drawing of a photograph obtained by this method is shown in Fig. 43. The two upper bands A represent the traces of the pencil of rays obtained by reversal of the magnetic field for the rays from the uncovered half of the wire; the lower bands B were obtained for the rays from the wire, when covered with eight layers of aluminium foil, each of thickness about .00031 cms. The apparatus was exhausted during the experiment, so that the absorption of the rays in air is negligible.

The greater deflection of the pencil of rays which have passed through the aluminium is clearly seen in Fig. 43. It will be shown later that the value of e/m for the particles does not change in consequence of their passage through matter. The greater deflection of the rays is then due to a decrease of their velocity after passing through the screen. This velocity is inversely proportional to the distance between the centres of the bands.

[1] Rutherford: Phil. Mag., July, 1905; Jan. and April, 1906.

We have seen that the α particles from radium C are all projected initially at the same speed. The absence of dispersion of the rays after passing through the screen shows that the velocity of all the α particles is reduced by the same amount in traversing the screen.

The following table gives the velocity of the α particles from radium C after passing through successive layers of aluminium foil, each of thickness about .00030 cms. The velocity is expressed in terms of V_0, the velocity of the α particles from radium C with an uncovered wire.

Number of layers of aluminium foil.	Velocity of α particles.
0	1.00 V_0
2	.94 "
4	.87 "
6	.80 "
8	.72 "
10	.63 "
12	.53 "
14	.43 "
14.5	not measurable

There is a marked weakening of the photographic effect of the rays after passing through 10 layers of foil. The photographic impression is weak, but distinct with 13 layers, and, using very active wires, can be observed with 14 layers. On account of this falling off of the photographic effect of the rays, very active wires are required to produce an appreciable darkening of the plate through more than 12 layers of foil. The lowest velocity of the α particle so far observed was about .4 V_0, which corresponded to the velocity of the rays after passing through 14 layers of foil. The photographic action of the α rays steadily diminishes with increase of the absorbing layer, but falls off very rapidly for a thickness greater than 10 layers of aluminium. The velocity of the α particle measured in this way is still considerable when the photographic action has almost ceased. Such a result suggests that there is a critical velocity of the α particles, below which they are unable to affect appreciably a photographic plate.

A similar abrupt falling off is observed in the ionizing and phosphorescent effects of the rays. From observations on thin layers of radium, Bragg found that the ionizing action of the rays from radium C ceased comparatively abruptly after traversing 7.06 cms. of air. A similar result was observed later by McClung by using an active wire coated with radium C as a source of rays.

In a similar way the writer found that the scintillations produced by the *a* rays on a screen of zinc sulphide ceased suddenly when the rays passed through 6.8 cms. of air. If layers of aluminium foil are placed over the active wire, the range of ionization and phosphorescence is diminished by a definite amount for each layer. Each layer of foil of the thickness used in the photographic experiment was equivalent in stopping power to about .50 cms. of air. A photographic effect of the *a* rays was just observable through 14 layers of aluminium. This corresponds to 7.0 cms. of air, — nearly the same range at which the ionizing and phosphorescent effects vanish. The three characteristic actions of the *a* rays thus cease together when the rays have passed through a definite distance of air or a definite thickness of an absorbing screen. Unless the velocity of the *a* particle falls off with great rapidity at the end of its course in air, it would appear as if there were a critical velocity of the *a* particle below which it produced no appreciable ionizing, photographic, or scintillating effect. This property of the *a* particles will be discussed in more detail later. In any case the rapid falling off of these three actions of the *a* particle at the end of its range indicates that there is a close connection between them. The photographic action of the *a* particle falls off in the same rapid manner as the ionizing action, and it seems reasonable to suppose that the effect of the rays on a photographic plate is the result of the ionization of the silver salts.

In a similar way, it is possible that the scintillations observed in zinc sulphide are primarily caused by ionization of the substance, and that the scintillations may arise as a result of the recombination of such ions. The brightness of the scintillations certainly depends on the velocity of the *a* particle. If the effect of the α rays on zinc sulphide is, as some have supposed, purely mechanical, and the scintillations result from a cleavage of the crystals, it is not easy to see why this effect should fall off suddenly, although the energy possessed by the particle is still considerable.

Electrostatic Deflection of the α Rays

In order to measure the deflection of the α rays from radium C in an electric field, the arrangement shown in Fig. 44 was adopted.

The rays from the active wire W, after traversing a thin mica plate in the base of the brass vessel M, passed between two parallel insulated plates A and B, about 4 cms. high and 0.21 mm. apart. The distance between the plates was fixed by thin strips of mica placed between them. The terminals of a storage battery were connected with A and B, so that a strong electric field could be produced between the two plates. The pencil of rays after emerging from the plates fell on a photographic plate P placed a definite distance above the plates. By means of a mercury pump the vessel was exhausted to a low vacuum. In their passage between the charged plates, the α particles describe a parabolic path, and after emergence travel in a straight line to the photographic plate. By reversing the electric field the deflection of the pencil of rays is reversed.

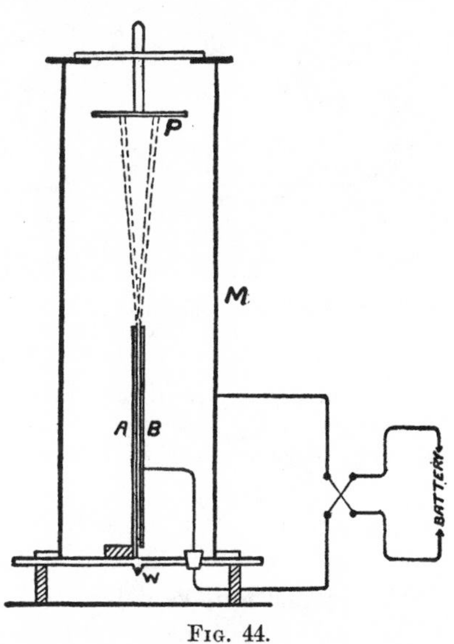

Fig. 44.

Apparatus for determining the deflection of α rays in passing through a strong electric field.

In Fig. 45 A shows the natural width of the line on the plates when no electric field is acting; B and C show the traces of the deflected pencils of rays for potential differences between the plates of 340 and 497 volts respectively. For small voltages the natural width of the line is broadened; for increased voltages the single line breaks into two and the width of the lines steadily narrows. Such an effect is to be expected theoretically. It can be easily shown that if D is the distance between the extreme edges of the deflected band for a potential difference E,

$$\frac{m v^2}{e} = \frac{8 E l^2}{(D - d)^2},$$

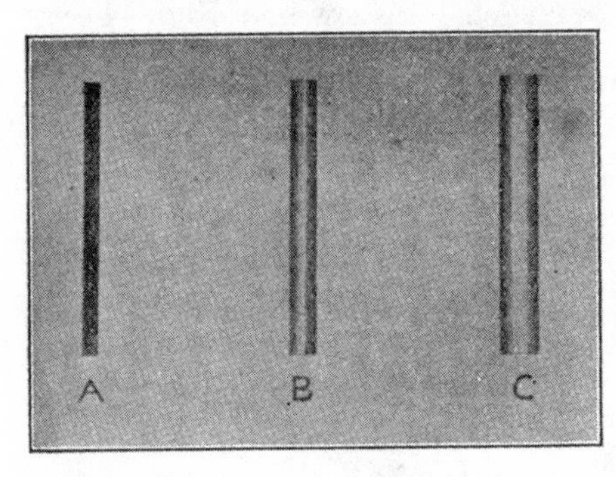

Fig. 45.

Electrostatic deflection of the α rays. The bands are drawn to scale from the actual photographs. Magnification about 3 times.

where e is the charge on the α particle, m its mass, v its velocity, l the distance of the photographic plate from the end of the parallel plates, d the distance between the parallel plates. This simple equation holds only if the field is strong enough to deflect the α particle through a distance greater than d in its passage through the electric field. For small values of the field, a modified form of this equation must be used.

The decrease in velocity of the rays in passing through the mica screen was separately determined. In most of the experiments, the mica plate reduced the velocity of the α particles from radium C by 24 per cent.

From the magnetic deflection the value of $\frac{m v}{e}$ is known, while from the electrostatic deflection, the value of $\frac{m v^2}{e}$ is determined.

From these two equations the values of e/m and v can at once be deduced. Examined in this way, it was found that:[1]—

[1] Rutherford: Phys. Review, Feb., 1906.

(1) The value of e/m was unaltered by the passage of the α particles through matter;

(2) The value of e/m was very nearly 5×10^3;

(3) The initial velocity of projection of the α particles from radium *C* was 2×10^9 cms. per second.

A similar method was applied to determine the value of e/m and the velocity of the α particles emitted from radium A and radium F (radiotellurium). In both cases the value of e/m was 5×10^3 within the limits of experimental error. The initial velocity of the α particles from radium A was about 86 per cent of that of the α particle from radium C, while the velocity of the α particle from radiotellurium was about 80 per cent of that of the α particle from radium C. The experiments on the velocity and value of e/m for the α particles from radium itself and from the emanation are not yet fully completed, but the results so far obtained indicate that the value of e/m will be the same as in the other cases.

Such results show conclusively that the α particles from radium and its products have identical mass, but differ in the initial velocities of their projection. The arguments in favor of the view that the α particle consists of an atom of helium carrying two ionic charges have already been discussed in detail on page 184.

Dr. Hahn, working in the laboratory of the writer, has found that the α rays emitted by thorium B are deflected both in a magnetic and in an electric field. These rays have a velocity about 10 per cent greater than those from radium C, but have the same value of e/m. In these experiments, the thorium B was deposited on a thin negatively charged wire, by exposure to the thorium emanation emitted by the very active preparation of radiothorium separated by Hahn (see page 68). It was also found that the range in air of the α particles expelled from thorium B, determined both by the electrical and scintillation methods, was about 8.6 cms., or about 1.6 cms. greater than that for the α particles from radium C.

Since the mass of the α particle from thorium B and the radium products is the same, it appears probable that the same equality also holds for the α particles from the other thorium products. The mass of the α particles from actinium has not yet been measured, but there is every reason to believe that it has the same value as for radium. On this view, the only common product of the different radioactive bodies is the α particle, which, as we have seen, is a projected helium atom.

Scattering of the α Rays

It is well known that a narrow incident beam of β or cathode rays is scattered in its passage through matter, so that the emergent pencil of rays is no longer well defined. This scattering of the β rays increases as the velocity of the β particles diminishes. In a theoretical paper, Bragg[1] pointed out that this scattering of the β rays is to be expected. The β particle in its passage through the molecules of matter enters the electric field of the atom, and its direction of motion is consequently changed. The smaller the kinetic energy of the β particle the greater will be the deflection of the path of some of the rays. If a narrow pencil of β rays falls on an absorbing screen, a portion of the rays will suffer so much deflection that the emerging beam will consist of a much wider cone of rays.

On account of their much greater kinetic energy, it is theoretically to be expected that the α particles will suffer much less deflection of their path in their passage through matter than the β rays. The α particles must move nearly in a straight line, and pass directly through the atoms or molecules of matter in their path, without much change in their direction of motion. This theoretical conclusion of Bragg is borne out by experiment. The scattering of the α particles is very small compared with that of the β particles moving at the same speed, so that a narrow pencil of α rays after traversing an absorbing screen is still well defined after emergence. At the same time there is undoubtedly a small scattering of the rays in their passage through matter, which must be taken into account.

If the α rays pass through air, for example, the width of the trace of a pencil of α rays on a photographic plate is always

[1] Bragg: Phil. Mag., Dec., 1904.

broader than in a vacuum. In addition the edges of the bands are not nearly so well defined in air as in a vacuum. This result shows that some of the α particles have suffered a change of direction of motion in their passage through the molecules of air.

The arrangement adopted to determine the retardation of the velocity of the α particles in their passage through matter (Fig. 43) is not complicated by the scattering of the rays, since the absorbing screen is placed over the active wire between the source and the slit. If, however, the absorbing screen is placed over the slit, the scattering of the α particles is at once seen by the broadening of the trace of the rays on the plate. A broad diffuse impression is observed on the plate instead of the narrow band with well defined edges, observed when the absorbing screen is placed below the slit. The amount of scattering increases with the thickness of the screen. When eleven layers of aluminium foil were placed over the slit, — an amount nearly sufficient to cut off the ionizing and photographic effects of the rays — an examination of the photographic plate showed that some of the rays had been deflected about 3° from the normal. A part of the rays may have been deflected through a considerably greater angle, but their photographic action was too small to be detected.

We may thus conclude that the path of the α particles, especially when their velocity is reduced, is deflected to an appreciable extent in passing through matter. The fact that the direction of motion of an α particle possessing such great energy of motion, can be changed in its passage through matter, shows that there must exist a very strong electric field within the atom, or in its immediate neighborhood. The change of direction of 3° in the direction of motion of the particles in passing through a distance of .003 cms. of matter would require an average transverse electric field over this distance corresponding to more than 20 million volts per centimetre. Such a result shows out clearly that the atom must be the seat of very intense electrical forces — a deduction in harmony with the electronic theory of matter.

We have seen that the velocity of the α particles from radium C lose their photographic action when their velocity falls to about 40 per cent of the initial value. On account of the complications introduced by the scattering of the a particles in passing through matter, it is difficult to decide with certainty whether this "critical velocity" of the a particle, below which it fails to produce its characteristic effects, is a real or only an apparent property of the rays. Without discussing the evidence in detail, I think there is undoubted proof that this critical velocity of the a particle has a real existence.

Photographic Effects of the a Rays from a Thick Layer of Radium

Since the a particles emitted by radium and its products decrease in velocity in their passage through matter, the radiations emerging from a thick layer must consist of particles moving at widely different speeds. This must obviously be the case, since the a particles which come from some depth below the radiating surface are retarded in their passage through the radium itself.

A pencil of rays from radium is consequently complex, and if a magnetic field is applied perpendicularly to the direction of the rays, each particle will describe the arc of a circle, the radius of which is directly proportional to the velocity of the particle.

This unequal deflection of the a particles in a magnetic field gives rise to a "magnetic spectrum," in which the natural width of the trace is much increased. This dispersion of the complex pencil of rays has been observed by Mackenzie[1] and by the writer.[2]

We have seen that the a particle has comparatively little photographic effect when its velocity falls to about .6 V_0, where V_0 is the maximum velocity of the a particle from radium C. Since the a particles from the latter have a greater speed than the a particles from any other radium product, we might thus expect to obtain a magnetic spectrum corresponding to a particles whose velocity lies between .6 V_0 and V_0. In an actual photograph of a deflected pencil of rays, the writer observed the

1 Mackenzie: Phil. Mag., Nov., 1905.
2 Rutherford: Phil. Mag., Jan., 1906.

presence of rays whose velocities lay between .67 V_0 and .95 V_0; while Mackenzie, by the method of scintillations, observed the presence of rays having velocities between .65 V_0 and .98 V_0.

When we remember that the photographic action of the β and γ rays from the radium prevents the detection of weak photographic effects produced by the α rays, the observations are seen to be in good agreement with theory.

Becquerel[1] early observed an interesting peculiarity in the deflection of a pencil of α rays from a thick layer of radium in passing through a uniform magnetic field. A narrow vertical pencil of rays fell on a photographic plate which was placed at right angles to the slit and inclined at a small angle with the vertical. By reversing the magnetic field, two fine diverging lines S P, S P′, were observed on the plate (see Fig. 46). The distance between these lines at any point represents twice the deflection of the pencil of rays from the normal at that point. By careful measurement, Becquerel found that these two diverging lines were not accurately the arcs of a circle, but that the radius of curvature of the path of the rays increased with the distance from the source. Becquerel thought that the α rays from radium were homogeneous, and concluded from this experiment that the value of e/m of the particles progressively decreased in their passage through air, in consequence of an increase of m by accretions from the air.

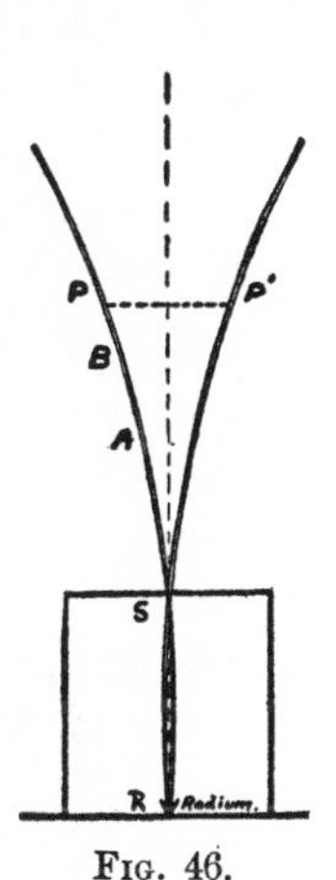

Fig. 46.

Bragg,[2] however, showed that this peculiarity in the trace of the rays could be simply explained without any assumption of an alteration in the value of e/m, by taking into account the complexity of the pencil of rays. The experimental arrangement is diagrammatically shown in Fig. 46. S P and S P′ represent the diverging traces of the rays on the photographic plate in a uniform magnetic field after emerging from the slit S. Let us

[1] Becquerel: Comptes rendus, cxxxvi, pp. 199, 431, 977, 1517 (1903).

[2] Bragg: Phil. Mag., Dec., 1904; April, 1905.

consider, for example, the outside edge of the trace at a point A. The photographic effect at this edge of the trace is due to the particles of lowest velocity, which are just able to produce photographic action at A. Consider next a point B further removed from the source. The α particles, which produce the edge of the trace, have the same velocity as in the first case; but since they have had to travel through a distance B R of air instead of A R, they must have initially started with a greater velocity. This must evidently be the case, since the α particle is retarded in its passage through air. The average velocity of these α particles along their path is consequently greater than in the first case, and the outside edge will be deflected through a smaller distance than would be expected if the average velocity were the same for the two paths A R, B R. This will cause the trace of the rays to show evidence of steadily increasing radius of curvature as we proceed from the source, — a result in agreement with the observations of Becquerel.

Quite a contrary effect is produced on the inside edge of the trace, for this is produced by the swiftest α particles from radium, viz., those emitted from radium C. Since the velocity of these particles decreases in their passage through air, the inside edge will show evidence of decreasing radius of curvature. This will have the effect of contracting the natural width of the trace. This effect is, however, small, and would tend to be masked experimentally by the scattering of the rays in their passage through air.

There is another paradoxical effect exhibited by a complex pencil of radium rays. Becquerel[1] showed that the outside edge of the trace of rays obtained in a magnetic field is unaltered by placing absorbing screens over the radium. In the case of a homogeneous source of α rays, we have seen that the pencil of rays suffers a greater deflection after passing through an absorbing screen. The absence of this effect in a complex pencil of rays from radium led Becquerel to believe that the α particles from it did not decrease in velocity in their passage through matter. We have seen, however, that for each indi-

[1] Becquerel: Comptes rendus, cxli, p. 485 (1905); cxlii, p. 365 (1906).

vidual product of radium, the α particles do suffer a retardation of velocity under such conditions.

The explanation of this apparent paradox is simple. The outside edge of the trace of the complex pencil of rays is due to the lowest velocity α particles which are just able to produce an appreciable photographic effect. The velocity of these α particles has been seen to be in the neighborhood of .6 V_0, where V_0 is the initial velocity of projection of the particles from radium C. When an absorbing screen is placed over the radium all the α particles suffer a retardation of velocity. The outside edge of the trace of the rays is produced by α particles of the same velocity as before; not, however, by the same α particles, but by another set, whose velocity has been diminished to this minimum amount in their passage through the absorbing screens.

The absence of increased deflection of the pencil of rays by the addition of absorbing screens is thus to be expected. These anomalies in the behavior of a complex pencil of rays show how necessary it is to use a source of homogeneous rays for the investigation of the properties of the α rays.

An account of these peculiarities of a complex pencil of α rays has been given in some detail, partly because of their great interest, and partly because the explanation of the effects has been a subject of some discussion.

Absorption of the α Rays

It was early recognized that the α particles were stopped in their passage through a few centimetres of air or by a few thicknesses of metal foil. On account of the weak ionization produced by uranium and thorium, it was not at first possible to work with narrow cones of α rays; but experiments were made with a large area of radioactive matter spread uniformly over a plate. The saturation current was measured between this plate and another plate placed parallel to it at a distance of several centimetres. As successive layers of aluminium or other metal foil were placed over the active matter, the ionization current was found to fall off approximately according to an exponential law with the thickness of the screen. A thick layer of radioactive matter was generally employed, and in the case of radium the exponential law appeared to hold fairly accurately over a considerable range.

Some experiments on the absorption of the *a* rays from an active preparation of polonium were made in a different way by Mme. Curie. The rays from the polonium passed through a hole in a metal plate covered with a wire gauze, and the ionization current was measured between this plate and a parallel insulated plate placed 3 cms. above it. No appreciable current was observed when the polonium was 4 cms. below the aluminium window, but as this distance was diminished, the current increased very rapidly, so that for a small variation of distance there was a large alteration in the ionization current.

This rapid increase of the current indicated that the ionizing property of the *a* rays ceased suddenly after traversing a definite distance in air. By adding a layer of foil over the polonium this critical distance was diminished.

The observed fact that the ionization current between two parallel plates varied approximately according to an exponential law with the thickness of the absorbing screen, when thick layers of radioactive matter were employed, tended to obscure the true law of the absorption of the *a* rays; for an exponential law of absorption had been observed by Lenard for the cathode rays, and also in some cases for the X-rays. In 1904, the question was attacked by Bragg and Kleeman,[1] both on the theoretical and experimental side, and the interesting experiments made by them have thrown a great deal of fresh light on the nature of the *a* rays and on the laws of their absorption by matter.

In order to account for their experimental results, Bragg formulated a very simple theory of the absorption of the *a* rays. On this theory all the *a* particles from a thin layer of radioactive matter of one kind were supposed to be projected with equal velocities and to pass through a definite distance in air before absorption. The velocity of the *a* particle decreased in its passage through air in consequence of the expenditure of its kinetic energy in ionizing the gas. As a first approximation, it

[1] Bragg and Kleeman: Phil. Mag., Dec., 1904; Sept., 1905.

was supposed that the ionization produced by a single a particle per centimetre of its path was constant for a certain range, and then fell off very abruptly after the particle had traversed a definite distance of air. This "range" of the a particle varied for each a ray product on account of the differences in the initial velocities of the a particles expelled from the separate products. If an absorbing screen were placed in the path of the rays, the velocities of the a particles from a simple product were all diminished in a definite ratio, and the range in air of the emerging particles was reduced by an amount proportional to the thickness of the screen and its density as compared with air.

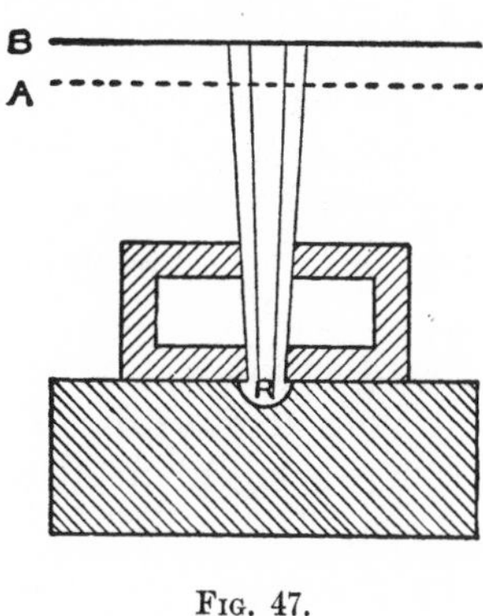

Fig. 47.

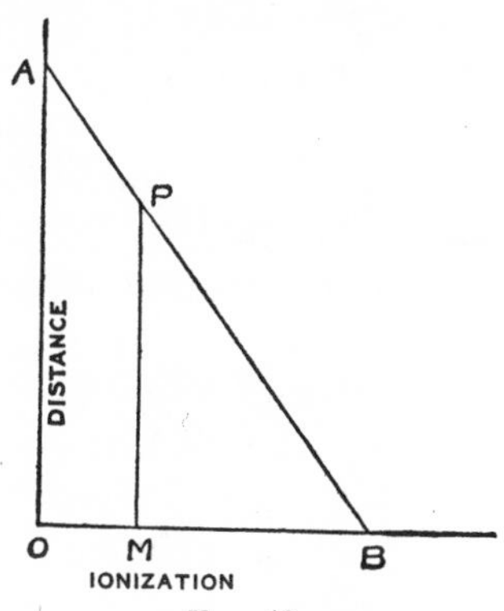

Fig. 48.

In a thick layer of radioactive matter, containing only one simple source of rays, the rays from the surface will have the maximum range a. Those emerging from a depth d of the radioactive matter of density ρ compared with air, will have a range in air of $a-\rho\, d$. With a thick layer of radioactive matter, the a particles emitted into the gas will vary widely in velocity and will have all ranges in air between zero and the maximum range a.

Suppose that a narrow pencil of a rays from a simple type of radioactive matter R (Fig. 47) passes into the ionization vessel A B through a wire gauze A. If the layer of active matter of one kind is so thin that the a rays are not appreciably retarded in their passage normally through it, the ionization at different distances from the source is expressed graphically in Fig. 48 by the curve A P M. The ordinates represent distance from the source of radiation and the abscissae the ionization produced in the vessel. The ionization commences suddenly at A and reaches a maximum at P, when the rays pass to the upper plate B of the ionization chamber and then remains constant till the source is reached.

With a thick layer, however, the rays have all ranges in air between the maximum and zero, and as the ionization vessel approaches the source, more and more of these *a* particles pass into it. The ionization curve is consequently then represented by a straight line A P B.

Theoretically, in order to obtain such results a narrow cone of rays and a shallow ionization vessel must be used. If the ionization vessel includes the whole narrow cone of rays, at all distances, the falling off of the intensity of the radiation according to the inverse square law need not be taken into account. The experiments of Bragg and Kleeman show that these theoretical conclusions are approximately realized in practice.

First, let us consider a thin layer of radioactive matter of one kind. This was obtained by evaporating a small quantity of radium bromide solution in a vessel. The emanation is driven off and the active deposit is transformed *in situ*. After about three hours the activity is due only to the *a* rays from the radium itself. The ionization curve obtained by Bragg and Kleeman is shown in Fig. 49, curve A. When the ionization chamber is more than 3.5 cms. above the source only a slight current is observed. At 3.5 cms. the current increases very rapidly and reaches a maximum at 2.85 cms. It then slowly falls off with decreasing distance. The maximum range of the *a* rays from radium itself is consequently 3.5 cms.

The corresponding curve for radium C is shown in the same figure, curve B. This was examined by McClung,[1] using the methods employed by Bragg and Kleeman. The radium C was deposited as an extremely thin film on a wire by exposure to the radium emanation. The rays from radium C had a maxi-

[1] McClung: Phil. Mag., Jan., 1906.

mum range of about 6.8 cms, and the ionization fell off in a very similar way to that observed by Bragg for the radium rays.

In Bragg's experiment the ionization chamber had a depth of 2 mms., while in the experiments of McClung the depth was 5 mms. In the case of radium C the ionization is seen to be nearly uniform for a distance of about 4 cms., and then to increase rapidly, the maximum ionization being reached at a distance of 5.7 cms. Allowing for the fact that the ionization chamber had a sensible depth, and that a fairly wide cone of rays was employed, it can be shown that the ionization must increase rapidly at a distance of 6.8 cms., but not quite so rapidly as the simple theory supposes.

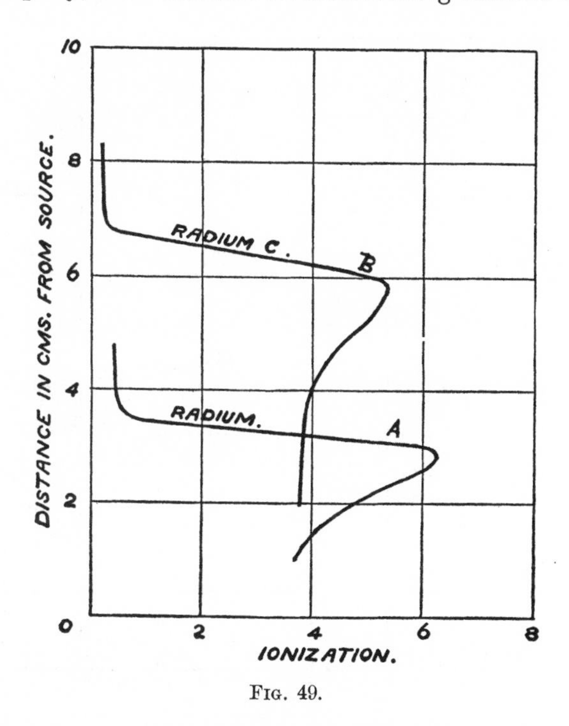

FIG. 49.

From a comparison of the diminution of velocity of the a particle from radium C in passing through aluminium, it can be readily calculated that the velocity of the a particles at the elbow of the curve is about .56 of the initial velocity of projection. At this velocity the a particle appears to be most efficient as an ionizer.

Bragg and Kleeman have examined by this method the range of the rays of the different a ray products present in radium in radioactive equilibrium. A thin layer of radium was employed, and the ionization curve is shown in Fig. 50.

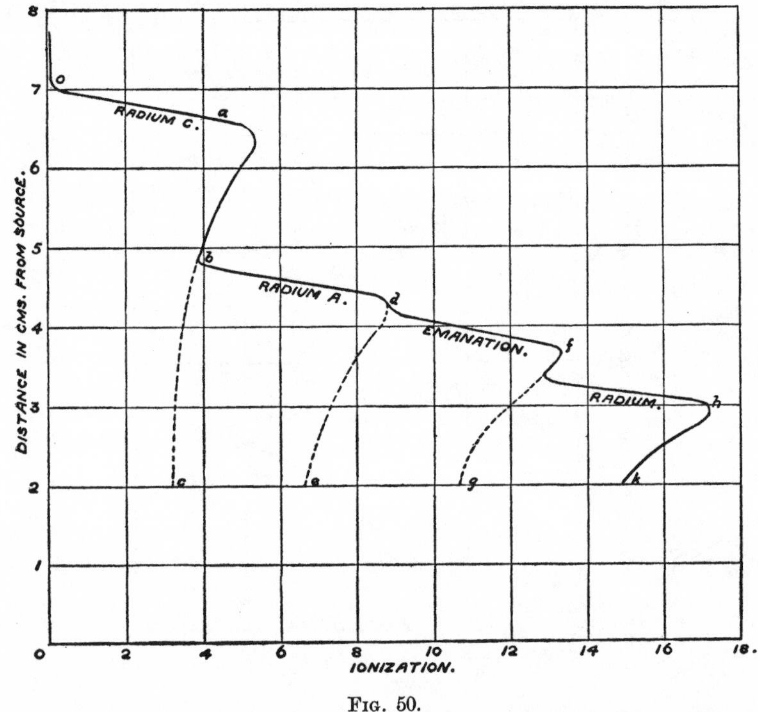

FIG. 50.

The first a rays entered the testing vessel at a distance of 7.06 cms. from the source. These rays were emitted from radium C, and have the greatest range of all the rays from the radium products. At a point b the curve suddenly turns through an angle, showing that at this point the a rays from another product, whose range in air is 4.83 cms., have entered the testing vessel. There is a similar though not so well defined break in the curve at *d* for a distance 4.23 cms., showing that another set of rays has entered the testing vessel. The break at *f* is due to the appearance of the rays from radium itself in the vessel. We may conclude from these results that the α particles from radium have a range of 3.5 cms. in air, and those from radium C a range of 7.06 cms. The ranges 4.23 and 4.83 cms. belong to the emanation and to radium A, but on account of the rapid change of A it has not yet been found possible to decide which of these two numbers belongs to the rays from the emanation and which to those from radium A.

If the curve *o a b* is produced downwards to *c*, the curve *o a b c* represents the ionization due to radium C alone at different distances from the source. Let this curve be now added to itself, being first lowered through a distance 2.23 cms., corresponding to the difference in range between 7.06 cms. and 4.83 cms. The new curve *b d e* lies accurately along the experimental curve *b d*. If the curve be again lowered through a distance 6.0 mms, corresponding to the difference of range for the next products, and a similar addition be performed, the resulting curve *d f g* again lies on the experimental curve. Finally if the curve is lowered through 7.3 mms., it is similarly found that the theoretical curve lies on the experimental curve *f h k*.

Knowing the ionization curve of one product, the experimental curve for the combined products can thus be built up from it in a very simple way. Such a result shows clearly that, allowing for the differences in the initial velocities of projection, the ionization curves for radium and each of its products are identical. It also shows that the same number of α particles are projected per second from each of the α ray products. This result follows from the disintegration theory if the various products are successive.

The results of Bragg and Kleeman have thus confirmed in a novel and striking way the theory of successive changes, initially developed from quite distinct considerations. They show that the products are successive, for otherwise the experimental curve could not be built up from consideration of the ionization curve of one product alone.

We may thus conclude from this evidence that radium A and C are true successive products, although it is difficult to test this point satisfactorily by direct experiment. The results also indicate that the α particles from all products are identical in all respects except velocity — a result confirmed, as we have seen, by direct measurements.

The method developed by Bragg and Kleeman thus not only throws light on the nature of the absorption of α rays, but indirectly affords a powerful means of determining the number of α ray products in radioactive matter, even if chemical methods should fail to isolate these products from the parent substance. This is possible if the α particles from the separate products have different ranges in air. A series of breaks in the ionization curve is a direct indication of the presence of a number of distinct radioactive substances which emit α rays. By this method, Dr. Hahn has shown that thorium B, which was supposed to contain one product, in reality contains two. From the difficulty of separation of these two products by chemical or physical methods, it appears probable that one has an extremely rapid period of transformation.

We have so far considered the ionization curves for a thin layer of radium, as this brings out the essential features of the absorption of α rays with great clearness. Bragg and Kleeman have also determined the ionization curves for a thick layer of radium. The curve is shown in Fig. 51. The curve consists of a number of straight lines meeting at fairly sharp angles. Above Q the ionization is due to the rays from radium C. At Q the α particles from the product of range about 4.8 cms. enter the ionization chamber, and the curve starts off at a sharp angle. A similar break is observed at R and S when the α particles from the other two products enter the ionization chamber. The slopes of the curves P Q, Q R, R S, S T, are very nearly in the ratio of 1, 2, 3, and 4, — a result to be expected from the simple theory.

Experiments were also made by Bragg and Kleeman on the absorption of the α rays by thin metallic layers and by other gases besides air. The effect of placing a uniform absorbing screen over a thin layer of radioactive matter is to depress the ionization curve by the same amount throughout its whole course. For example, the loss of range for an absorbing screen of silver foil, whose weight per unit area was .00967 grams, was equal to that for a stratum of air of thickness 3.35 cms. and

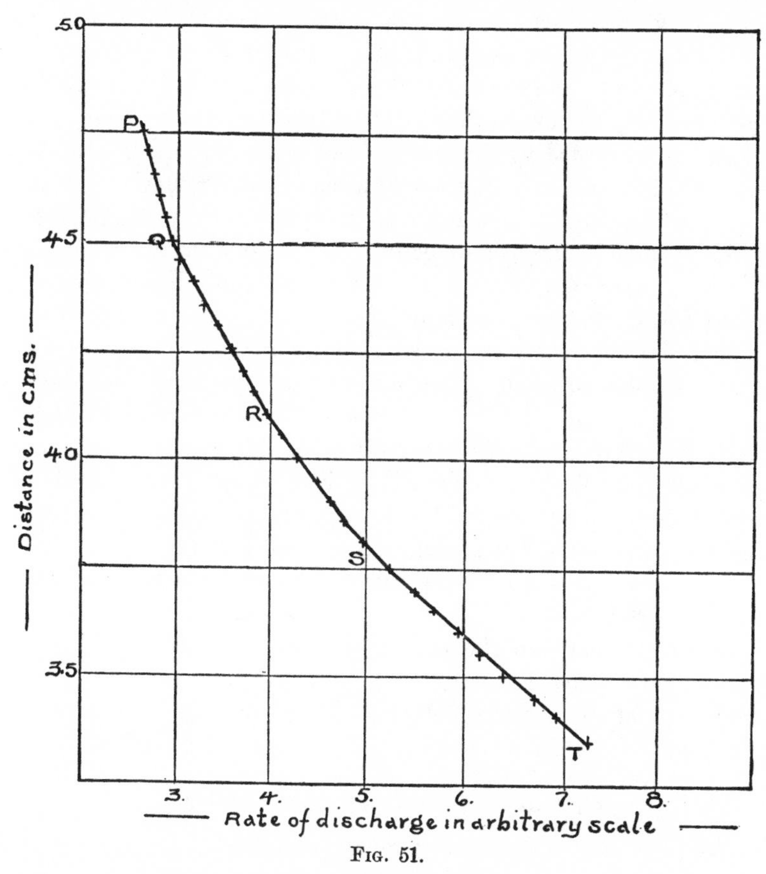

FIG. 51.

whose weight per unit area was .00402 gram. The ratio of these two weights is 2.41, showing that the stopping power of silver is 2.41 times greater than would be expected on a simple density law. An examination of a number of metals showed that the stopping power was approximately proportional to the square root of the atomic weights. A similar law was found to hold for gases over a considerable range of density. This relation is most remarkable, and indicates that the absorption of energy in the atom is proportional to the square root of its atomic weight. It is known that for simple gases like hydrogen, oxygen, and carbon dioxide, the total number of ions produced by complete absorption of α rays of given intensity is nearly the same, suggesting that the same energy is required in each case to produce an ion. If the stopping power of any gas is mainly governed by the energy used up in producing ions, the results obtained by Bragg and Kleeman indicate that on an average four times as many ions are produced by the passage of an α particle of given velocity through an atom of oxygen as in its passage through an atom of hydrogen. This does not necessarily assert that each atom of the gas in the path of the α particle is ionized, but is supposed to be an average result when a larger number of atoms is considered. At the same time, there is strong evidence that the number of ions produced by an α particle in air is at least as great as, if not greater than, the number of molecules with which it collides. We are thus driven to suppose either that the α particle is able to produce more than two ions out of each molecule of a heavy gas, or that the sphere of action of the α particle is greater in a dense than in a light gas.

Whatever may be the conclusions to be drawn from the experiments, they certainly serve to show that there is some fundamental connection between ionization and the atomic weight of different elements.

Charge carried by the α Rays

We have seen that the α particle is deflected in a magnetic or electric field as if it carried a positive charge of electricity. It was early observed that the β particles from radium carried with them a negative charge, and that the radium from which they were expelled gained a positive charge. This property of radium is illustrated in a striking manner in a simple apparatus devised by Strutt, known as the "radium clock." Two gold leaves are in metallic connection with an insulated tube containing radium, and the whole is placed in a vessel exhausted to a low vacuum. The β particles are fired through the radium tube, carrying with them a negative charge, and leave behind an equal positive charge. The leaves gradually diverge with positive electricity, and by a suitable contact are made to discharge automatically after a certain divergence has been reached. This process of charging and discharging continues indefinitely, or at least as long as the radium itself will last. Using 30 milligrams of radium bromide, the leaves may be made to pass through the cycles of charge and discharge several times a minute.

If a rod or plate, covered with a thin film of radium, which has been heated to get rid of the β and γ rays, is exposed in a similar way, no such charging action is observed, however good a vacuum is produced. If the insulated plate is charged either positively or negatively, the charge is rapidly lost.

Experiments of this kind are most simply made with a plate coated with a thin film of radiotellurium (radium F). This substance has the advantage of emitting α rays but not β rays. The reason for the failure to detect the charge carried by the α rays in the earlier experiments was made clear by an investigation of J. J. Thomson. He showed that such an active plate emitted in addition to the α particles a large number of slow moving electrons, which had very little power of penetration, and moved with so slow a velocity that they could readily be turned back or bent from their course by electric or magnetic fields. The presence of a large number of these negatively charged particles under ordinary conditions completely masks the charge carried by the α particles. The effect of these slow moving electrons, however, can be almost entirely eliminated by applying a strong magnetic field parallel to the plane of the active plate. The electrons emitted from the plate then describe a curved path in the magnetic field and return to the plate from which they set out. Under such conditions, in a highly exhausted vessel, it can be shown that the plate acquires a negative charge, while a body on which the α particles impinge receives a positive charge.

Such results show clearly that the a particles are expelled with a positive charge, but that they are always accompanied by a large number of slow moving electrons. These electrons appear to be a type of secondary radiation set up by the escape of the a particles from the active matter and by the matter on which they impinge. Their presence has been noted not only in radiotellurium but also in radium itself, in its emanation, and in the emanation from thorium. These electrons appear to be a necessary accompaniment of the emission of a particles, but must not be confused with the β rays proper, which are projected at a much greater speed and have a much greater penetrating power. By employing a magnetic field to get rid of the disturbance caused by the slow moving electrons, the writer determined the charge carried by the a rays from a thin film of radium spread uniformly on a metal plate. Knowing the quantity of radium on the plate, it was deduced that 6.2×10^{10} a particles were projected per second from a gram of radium at its minimum activity. For radium in equilibrium, which contains four a ray products, the corresponding number is 2.5×10^{11}. These calculations are based on the assumption that each a particle carries a single ionic charge of value 3.4×10^{10} electrostatic units. If the a particle carries double this charge the number expelled is only one half of the above.

By measurements of the charge carried by the a rays from radium C, obtained on a lead rod by exposure to the radium emanation, it was calculated that 7.3×10^{10} β particles are expelled per second from one gram of radium. Recently Schmidt has shown that the supposed rayless product radium B, as well as radium C, emits β particles, but that these have much smaller penetrating power. If equal numbers of β particles are expelled per second from radium B and C the number for each β ray product per gram of radium is 3.6×10^{10}.

McClelland has shown that a strong secondary radiation is set up by the impact of β particles on lead. It is thus probable that the number 3.6×10^{10} is too high, for the β particles fired into the lead give rise to secondary β particles whose charge is measured with that of the primary β particles. If each a particle carries twice the charge of the β particle, the number of β particles expelled per second for each product in a gram of radium should be 3.1×10^{10}. Although it is difficult to draw any very definite conclusions from such comparisons, the evidence is in agreement with the view that in the product radium C which emits α and β rays, the number of α and β particles emitted per second is the same, while the charge carried by the α particle is twice that carried by the β particle.

Heating Effect of the α Rays

In 1903, Curie and Laborde[1] made the striking discovery that radium was always hotter than the surrounding medium, and radiated heat at a constant rate of about 100 gram calories per hour per gram. The question immediately arose as to whether this phenomenon involved some new scientific principle, or whether it was merely a secondary effect due to the bombardment of radium by its own α particles.

Since the α particles have large kinetic energy and are very easily stopped by matter, most of those produced within the radium do not emerge, but are stopped by the radium itself, and their energy of motion is transformed into heat *in situ.* In measurements of its heating effect the radium is enclosed in a vessel of sufficient thickness to absorb all the α rays emitted from the surface. It is consequently not necessary to make any correction for the α particles which escape from the radium itself. It thus appeared probable that the heating effect of radium might result largely from the bombardment of the radium itself by the α particles produced within it.

Rutherford and Barnes[2] made a number of experiments to throw light on this subject. The heating effect of about 30 milligrams of radium bromide was first measured in a simple form of air calorimeter and was found to correspond to about 100 gram calories per hour per gram. The radium was then heated to a sufficient temperature to drive off the emanation, which was then condensed in a small glass tube by means of liquid

[1] Curie and Laborde: Comptes rendus, cxxxvi, p. 673 (1904).
[2] Rutherford and Barnes: Phil. Mag., Feb., 1904.

air, and the tube sealed. The variation in the heating effect of the radium so treated and of the emanation tube were then separately examined. After removal of the emanation, the heating effect of the radium fell rapidly in the course of about three hours to 27 per cent of its maximum, and then slowly increased again, finally reaching its old value after a month's interval.

The heating effect of the emanation tube varied in exactly the opposite way, for it increased to a maximum in about three

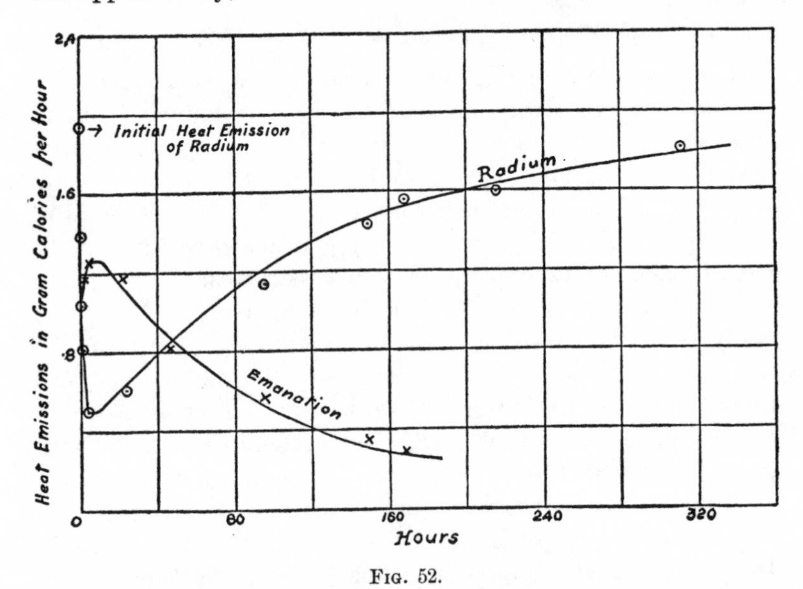

Fig. 52.

Variation of heating effect of radium after removal of its emanation and variation of that of the emanation and its products.

hours, when it was equal to about 73 per cent of the heating effect of the radium. It then gradually decreased according to an exponential law, falling to half value in about four days. The curves of recovery of the heating effect of the radium and of the loss of heating effect of the emanation tube are shown in Fig. 52. Within the limit of experimental error, the sum of the heating effects of the radium and emanation tube was equal to that of the radium in radioactive equilibrium. Allowing for the fact that 6 per cent of the emanation was not removed by the heating, it is seen that only 23 per cent of the heating effect is due to radium itself, and the other 77 per cent to the emanation and its products.

The decay and recovery curves of the heating effect are identical within the limit of experimental error with the corresponding curves of decay and recovery of activity measured by the α rays. Such a result indicated that the heating effect is a measure of the kinetic energy of the α particles, for radium has an α

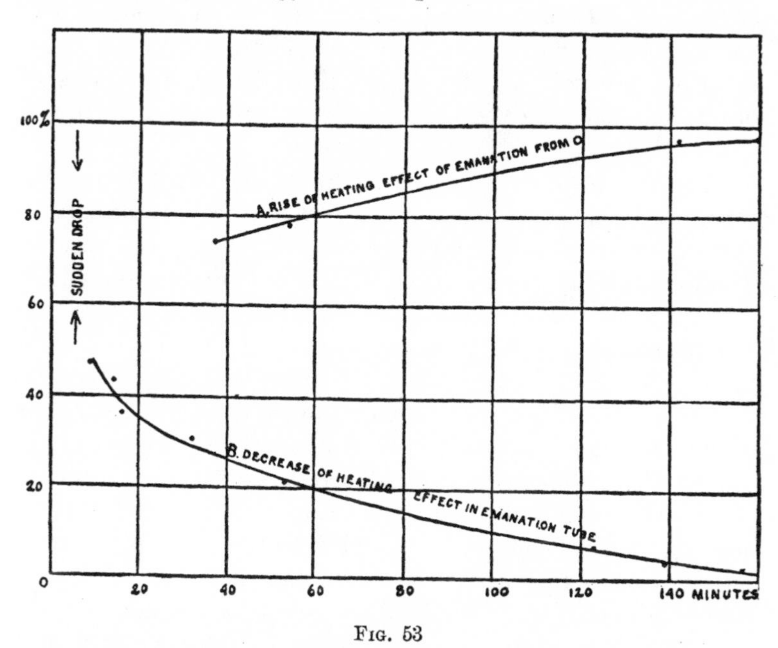

FIG. 53

ray activity of about 25 per cent of its maximum when the emanation and its products are removed, while the β and γ ray activity practically disappears. In order to test this point still further, the distribution of the total heating effect of the emanation tube between the emanation and its further products was determined. After measuring the heating effect of the emanation tube, the end of the tube was broken and the emanation was entirely removed. Ten minutes later the heating effect of the tube had dropped to 48 per cent, and it then diminished steadily to zero. The curve of decrease of heating effect is shown in Fig. 53. After ten minutes the curve follows the activity decay curve fairly closely. After removal of the emanation the heating effect of the tube is due to the active deposit consisting of radium A, B, and C, which remains behind. Since A loses its activity according to a three minute period, it is not possible to follow the variation of its heating effect. After fifteen minutes the heating effect must be due entirely to radium B and C. It is not easy to decide experimentally whether the rayless product radium B supplies an appreciable proportion of the heating effect, but from the absence of α rays, its heating effect is probably very small compared with that of C.

The curve of increase of heating effect (Fig. 53) of a tube into which the emanation has been introduced is complementary to that of the curve of decrease. Such a relation is to be expected. The fact that the heating effect falls off according to the period of each α ray product shows that the heat emission of radium and its products is mainly a consequence of the expulsion of α rays.

It was deduced from the experiments that about 23 per cent of the heating effect was due to radium alone, 32 per cent to radium C, and 45 per cent to the emanation and radium A together. On account of the rapid rate of the change of A, its heating effect cannot be disentangled from that of the emanation. Direct experiment has shown that not more than one or two per cent of the heat emission of radium is due to the β or γ rays even when they are completely absorbed by a lead envelope.

We shall now consider the important question as to whether the energy of motion of the α particles ejected from radium and its products is sufficient to account for the heating effect observed. The kinetic energy of an α particle of mass m moving with a velocity v is $\frac{1}{2} m v^2$. The relative kinetic energies of the α ray particles from each of the α ray products can at once be determined if their relative velocities are known. These velocities have not yet been directly measured for the rays from each product, but they can readily be deduced from the ranges of the α particles in air. For example, the velocity of the α particle from radium itself is equal to that of the particle from radium C after the latter has passed through a distance of air equal to the difference of the ranges of the two α particles in air. This difference is 3.5 cms., which corresponds to 6.7 layers of aluminium foil of the same thickness as that employed in the experimental results tabulated on page 225. In this way, taking the kinetic energy of the α particle from radium C as 100, it can readily be deduced that the kinetic energies of the α particles of range 4.8 and 4.3 cms. are 74 and 69 respectively. For the α rays from radium itself of range 3.5 cms., the energy is 58. Since the products in radium are successive, the same number of α particles is expelled per second from each product. Taking the kinetic energy of the α particle as a comparative measure of its heating effect, it can at once be deduced from these numbers that 19 per cent of the total heating effect should be produced by radium itself, 48 per cent by radium A and the emanation together, and 33 per cent by radium C. The corresponding values obtained by direct measurement of the heating effect are 23, 45, and 32 per cent respectively. The agreement between theory and experiment is thus fairly good.

Now, on the assumption that the α particle carries an ionic charge of 1.13×10^{-20} electromagnetic units, it has been experimentally found that each of the α ray products present in one gram of radium product expels 6.2×10^{10} α particles per second. The kinetic energy of the α particle from radium C is $\frac{1}{2} m v^2$. Substituting the known values $e/m = 5 \times 10^3$, and $v = 2.0 \times 10^9$, and $e = 1.13 \times 10^{-20}$, the kinetic energy is seen to be 4.5×10^{-6} ergs. The value of the kinetic energy of radium C deduced in this way is independent of whether the α particle carries one or two ionic charges. The kinetic energy of the α particles of radium C expelled per second from one gram of radium is thus 2.79×10^5 ergs. The heating effect of the α particles from radium C per hour per gram of radium is consequently 24 gram calories. The amount experimentally observed is 32 gram calories per hour.

Considering the experimental difficulty of accurately determining the number of *a* particles expelled from radium per second, the agreement of experiment and calculation is remarkably good, and shows clearly that the greater part of the heat emission of radium is due to self-bombardment by its own *a* particles. It is possible that a small fraction of the heat emitted by radium may be due to the energy liberated in consequence of the rearrangement of the atom after the violent expulsion of the *a* particle, but it appears probable that this would be small compared with the energy of motion of the *a* particle itself.

The conclusion that the heating effect of radium and its products is a measure of the energy of the *a* particles produced by it applies also equally well to the other radioelements which emit *a* rays. We should thus expect thorium, uranium, and actinium to emit heat at a rate approximately proportional to their *a* ray activities. Pegram has examined this question for thorium and found evidence of a rate of heat emission about that to be expected from its activity compared with that of radium. Every substance that emits *a* particles must thus also emit heat at a rate proportional to the product of the number of *a* particles produced per second and the average kinetic energy of each *a* particle.

The enormous heating effect of the radium emanation compared with the quantity of matter involved has already been discussed on page 91. The rapidly changing products like the actinium, and thorium emanations, and radium A, must initially emit heat at an enormous rate compared with the quantity of matter present. For example, the actinium emanation which has a period of 3.9 seconds must, weight for weight, emit heat at about 800,000 times the rate of the radium emanation. The average duration of the time during which heat is emitted is correspondingly less.

Gases evolved from Radium

We have already seen that helium is produced in small amounts from radium. Giesel, Runge, and Bödlander observed that radium solutions produced a considerable quantity of hydrogen and oxygen. Ramsay and Soddy found that 50 milligrams of radium bromide in solution evolved about 0.5 c.c. of mixed gases per day. About 28.9 per cent of this consisted of oxygen and the rest of hydrogen. The hydrogen is thus present in a slight excess compared with that obtained by the decomposition of water, and no satisfactory explanation of this excess has yet been given. It may possibly be due to oxidation of the radium bromide to radium bromate. Ramsay showed that the radium emanation mixed with water produced hydrogen and oxygen, and that after explosion of the mixed gases no observable bubble of gas remained. This evolution of gas proceeds at a constant rate, and must be the result of the action of the α radiations in decomposing the water molecules. One gram of radium bromide in equilibrium would produce about 10 c.c. of hydrogen and oxygen per day. The energy required to dissociate the corresponding amount of water per day is about 20 gram calories, or less than 2 per cent of the total kinetic energy of the α particles produced by radium.

In order to generate 10 c.c. of hydrogen and oxygen per day by electrolysis a steady current of .00067 amperes is required. Now it is found experimentally that the maximum ionization current in air due to one gram of radium bromide in equilibrium, spread in a thin film, is .0013 amperes, or about twice the current required to produce the amount of hydrogen and oxygen observed.

In the experiments of Ramsay and Soddy some of the emanation left the solution and collected in the open space above it, and the rate of production of gases observed is probably less than if the emanation were all retained in the solution. In addition, the α rays not only decompose water but conversely cause the combination of hydrogen and oxygen to form water. Taking such factors into account, it certainly appears more than a coincidence that the ionization current in air due to the α rays from radium is of the right magnitude to account for the observed production of hydrogen and oxygen.

The gradual loss of energy of the α particles in their passage through a gas appears to be mainly due to the energy absorbed in the ionization of the gas. The fact that the stopping power of matter in general, whether solid, liquid, or gaseous, is proportional to the square root of the atomic weight, suggests that matter of all kinds is ionized by the passage of the *a* rays. It is thus to be expected that about the same total number of ions would be produced by the complete absorption of the *a* rays in water as are produced by absorption in air. The appearance of hydrogen and oxygen in the radium solution is undoubtedly mainly a result of the ionization of the water molecules by the *a* particles, and shows that the ionization consists in large part in an actual chemical dissociation of the water molecules. It has been generally supposed that ionization in simple gases like helium, hydrogen, and oxygen is due to the expulsion of an electron from the molecule. This may be the case, but in a complex molecule like that of water, the ionization by the *a* rays consists in, or at any rate results in, an actual chemical dissociation of water into hydrogen and oxygen. Whether this dissociating action is a property only of the *a* rays, or of all powerful ionizing agencies, cannot be discussed at this point, but the evidence certainly suggests that the ionization of complex substances by the *a* rays is very similar in character to ionization of solutions, and consists in part in a chemical dissociation of the substance.

There is considerable evidence that the *a* particles produce chemical action of various kinds. For example, the *a* rays convert oxygen into ozone, coagulate globulin, and produce chemical changes in barium platinocyanide.

Summary of Properties of the *a* Rays

(1) The *a* particles from radium, and probably from all radioactive substances, consist of positively charged atoms of matter projected with great velocity.

(2) The *a* particles from radium and its products all have the same mass, and are probably atoms of helium.

(3) Each product of radium expels *a* particles at a definite speed, which is characteristic of that product, but varies for different products.

(4) The ionizing, photographic, and phosphorescent actions of the α rays from a simple product all appear to cease abruptly when the velocity of the α particle falls below a certain critical speed.

(5) The velocity of projection of the α particles increases steadily for the successive products of radium, and is greatest for those emitted from radium C. The maximum velocity is about 2×10^9 cms. per second.

(6) The velocity of the α particle is diminished in its passage through matter.

(7) The α rays from a thin layer of any simple product are homogeneous, *i. e.*, they consist of α particles all projected with the same velocity. On account of the retardation of the α particles in passing through matter, the rays from a thick layer of any simple radioactive substance are complex, *i. e.*, they consist of α particles projected with velocities varying over a considerable range.

(8) The initial velocities of projection of the α particles from radium and its products lies between 10^9 and 2×10^9 cms. per second.

(9) The heating effect of radium is a result of the bombardment of the radium by its own α particles.

CHAPTER XI

PHYSICAL VIEW OF RADIOACTIVE PROCESSES

In the preceding chapters, the more important properties of the radioactive bodies have been considered, and it has been seen that the results obtained receive a satisfactory explanation on the view that the radioactive matter suffers spontaneous disintegration.

We shall now endeavor to outline in as concrete a manner as possible the processes which are believed to take place in the atoms of radioactive matter, and in the medium surrounding them. Such representations of the nature of the atom and of the processes occurring therein are, in the present state of our knowledge, somewhat speculative and imperfect, but they are nevertheless of the greatest assistance to the investigator in providing him with a working hypothesis of the structure of the atom. The behavior of such model atoms can be compared with that of the actual atoms of matter under investigation, and in this way it is possible to form gradually a clearer and more definite idea of the constitution of the atom.

Modern physical and chemical theories are all based on the assumption that matter is discontinuous and is made up of a number of discrete atoms. In each element the atoms are supposed to be all of the same mass and of the same constitution, but the atoms of the different elements show well marked differences in physical and chemical behavior. It has been quite incorrectly assumed by some that the study of radioactive phenomena has tended to cast doubt on atomic theories. Far from this being the case, such a study has materially strengthened, if, indeed, it has not given actual proof of, the atomic structure of matter.

Any one who has witnessed the multitude of scintillations produced on a zinc sulphide screen by the a rays of radium cannot fail to have been impressed with the idea that radium is sending out a shower of small particles. Such a view is confirmed by direct measurement, for we know that the scintillations are due to the α particles, which consist of minute material bodies, all of the same mass, which are expelled from the radium at an enormous speed. The energy of motion of each α particle is so great that in some cases the impact of the α particle on the screen is accompanied by a visible flash of light. These α particles, as we have seen, are not fragments of radium, but atoms of helium.

While the study of radioactivity has emphasized the ideas of the atomic constitution of matter, it has at the same time indicated that the atom is not an indivisible unit but a complex system of minute particles. In the case of the radioactive elements, some of the atoms become unstable and break up with explosive violence, expelling in the process a portion of their mass. Such views are rather an extension than a contradiction of the usual chemical theory which supposes that the chemical atom is the smallest combining unit of matter in ordinary chemical change. The atom may be the smallest combining unit, and still be a complex system which cannot be broken up by any physical or chemical agencies under our control.

In fact, the great emission of energy in radioactive changes shows clearly the reason why chemistry has failed to break up the atom. The forces which bind together the component parts of an atom are so great that an enormous concentration of energy would be required to break up the atom by the action of external agencies.

The complex structure of the atom is clearly indicated by an examination of its spectrum. Under the stimulus of heat or the electrical discharge, the atom has certain definite periods of vibration which are characteristic of each particular element. Even in the case of a light atom like hydrogen, the great number of different periods of vibration extending far into the ultraviolet shows that the atom must be a complex structure which is able to vibrate in a variety of ways. The modes of vibration of a hydrogen atom are exactly the same under all conditions, and are the same, for example, for free hydrogen in the sun and for hydrogen prepared by different chemical processes on the earth.

The unchanging character of the spectrum of the elements has been urged by some as an objection to the view that atoms suffer disintegration. This objection, however, does not appear to be a very weighty one, for the present theories of atomic disintegration suppose that there is not a gradual variation of the properties of the atoms as a whole, but a sudden disintegration of a minute fraction of the total number present, the rest remaining quite unchanged. For example, the spectrum of radium itself remains unaltered so long as any radium remains untransformed. Supposing, however, we were able to detect the spectrum of the products mixed with it, we should find the spectrum of radium in equilibrium to consist of the normal radium spectrum, plus the spectrum of each of its products superimposed upon it. Each product would have a definite and characteristic spectrum, differing for each product, and having no apparent connection with that of the parent element.

Development of the Electronic Theory of Matter

The laws of electrolysis discovered by Faraday indicated that each atom of hydrogen carried an invariable charge, whose value e could be approximately deduced from data on the mass of the atom. The oxygen atom always carried a charge $2e$ and an atom of gold $3e$, and, generally, the ions of different elements in solution carried charges which were integral multiples of the charge carried by the hydrogen atom. In no case was any atom found to carry a charge less than e. The idea thus arose that the charge carried by the hydrogen atom was the smallest unit of electricity, and was not capable of further subdivision. Such a conception was practically equivalent to an atomic theory of electricity.

Theories of atomic constitution were advanced in which it was supposed that the atom consisted of a number of charged ions in motion. Such theories, of which the most notable exponents were Larmor and Lorentz, were primarily advanced to explain the mechanism of atomic radiation. More physical definiteness was given to such theories by the discovery of J. J. Thomson that the cathode rays consisted of a flight of particles, whose apparent mass was only about 1/1000 that of the hydrogen atom. These "corpuscles" or "electrons" were found to be emitted from substances under a variety of conditions. Not only were they obtained by means of the electric discharge in a vacuum tube, but also from a white hot carbon filament and from a zinc or other metal plate exposed to the action of ultra-violet light. They were found to be spontaneously emitted from the radioactive bodies, with velocities in many cases much greater than those attained in a vacuum tube.

At the same time the discovery by Zeeman of the effect of a magnetic field on the period of the light vibrations indicated that the vibrating system consisted of negatively charged particles, whose mass was about the same as that of the electron set free in a vacuum tube. Such results indicated that the electron was a constituent of all matter, and escaped from it under a variety of conditions.

It was at first assumed as the simplest hypothesis that the electron consisted of a material particle of mass about 1/1000 that of the hydrogen atom, and carrying the same charge as a hydrogen atom in the electrolysis of water. Theory had long before shown that a moving charge possessed electrical mass in virtue of its motion. The theory indicated that this electrical mass should be constant for slow speeds, but should increase rapidly as the velocity of light is approached. In order to test this theory definitely it was necessary to determine the value of e/m for the electrons moving at velocities closely approaching that of light.

Radium proved to be an ideal source of electrons for such an experiment, since it expels a particles over a wide range of speed, the velocity of some of them being very nearly equal to that of light. As we have seen, Kaufmann measured the velocity and value of e/m for the electrons from radium, and definitely showed that the apparent mass of the electron increased with its velocity. By comparison of theory with experiment, it was found that the mass of the electron was purely electrical in origin, and that there was no necessity to assume that the charge was distributed over a material nucleus. We thus arrive at the remarkable conclusion that the particles of the cathode stream and the β particles of radium are not matter at all in the ordinary sense, but disembodied electrical charges whose motion confers on them the properties of ordinary mass. It has already been pointed out (page 11) that ordinary mass itself may possibly be explained purely as a result of electricity in motion. In order to account for the observed magnitude of the mass of the electron at different speeds, it is necessary to suppose that the electricity is distributed over a surface of minute area or throughout a minute volume.

Taking as the simplest assumption that this surface is spherical in shape, it is necessary to suppose that the radius of the sphere on which the charge is distributed is about 10^{-13} cms. Now, from a variety of considerations, it can be calculated that the radius of an atom is about 10^{-8} cms., or rather that the sphere of action of atomic forces extends from the centre of the atom over about this distance. We thus see that if an atom were magnified so as to be represented by a sphere of 100 metres radius, the radius of an electron would be only one millimetre. Consequently, if we suppose that the hydrogen atom consists of a thousand electrons free to move within the dimensions of the atom, the electrons would be so sparsely distributed that they would occupy rather than fill the space within the atom, and would only occasionally interfere with the freedom of one another's individual motion.

Most of the magnetic and electric field surrounding an electron in motion lies close to its surface. This must evidently be the case, since the magnitudes of both these forces diminish inversely as the square of the distance, and become comparatively small at a distance from the electron corresponding to a few radii. The forces due to an electron in motion are thus mostly confined within a sphere of radius about 10^{-12} cms. The direction or magnitude of the motion of any electron would not be sensibly disturbed by the presence of another unless it approached within this small limiting distance.

It has been found experimentally that an ion produced in gases by X-rays, or the rays from active bodies, carries a positive or negative charge equal in magnitude to 3.4×10^{-10} electrostatic units. The charge carried by an ion in gases is apparently the same for all gases, and does not vary, as in the case of electrolysis, with the valency of the atom. The charge carried by an ion has been shown to be identical with that carried by a hydrogen atom set free in the electrolysis of water.

Although the charge carried by an electron has not been directly measured, there is every reason to believe that it is identical with the charge carried by the negative ion in gases. The charge on an electron is supposed to be the smallest unit of electricity that takes part in the transfer of an electric current, whether in solids, liquids, or gases. There is one marked point of distinction between a positive ion and an electron. The electron in motion has an apparent mass of about 1/1000 that of the hydrogen atom, while the corresponding positive charge has never been found associated with a mass less than that of the hydrogen atom. This has led to the view that there is only one kind of electricity, viz., negative, which is associated with the electron, and that a positively charged body or ion is one which has been deprived of one or more of its normal complement of electrons.

Radiation from an Electron

An electron in motion produces a magnetic field whose intensity at any point, for velocities small compared with that of light, is proportional to its velocity. This magnetic field travels with the electron, and magnetic energy is consequently stored up in the medium surrounding it. The amount of this magnetic energy is proportional to the square of the velocity u of the electron, and can consequently be expressed in the form $\frac{1}{2} m u^2$. In this equation m represents the apparent or electrical mass of the electron, and is equal to $\frac{2e^2}{3a}$, where e is the charge and a the radius of the electron.

An electron moving uniformly in a straight line does not

radiate energy, but any change in its motion is accompanied by the dissipation of energy in the form of electromagnetic radiation, which travels out from the electron with the velocity of light. The rate of dissipation of radiant energy is proportional to the square of its acceleration, and consequently becomes large if an electron is suddenly set in motion or brought to rest. For example, the X-rays produced in a vacuum tube are believed to consist of the intense electromagnetic pulses which are set up by the sudden arrest of the cathode rays when they impinge on the anticathode.

An electron constrained to move in a circular orbit is a powerful radiator of energy, since its motion is always accelerated toward its centre. This necessary loss of energy from an accelerated electron has been one of the greatest difficulties met with in endeavoring to deduce the constitution of a stable atom. For the supposition that an atom consists of a number of positively and negatively charged particles in motion, held in equilibrium by their mutual forces, Larmor[1] has shown that the condition for no loss of energy by radiation is that the vector sum of the accelerations of all the component charged particles shall be permanently zero. If this condition is not fulfilled there will be a steady drain of internal energy from the atom in the form of electromagnetic radiation, and unless this is balanced in some way by the absorption of energy from outside, the atom must ultimately become unstable and break up into a new system.

There are thus two essential conditions which must be fulfilled for atoms to be permanent. The positively and negatively charged particles constituting the atom must be so arranged as to form a stable aggregate under their forces of attraction and repulsion, and at the same time their arrangement and motions must be such that no energy is radiated from the atom.

Since there is reason to believe that the atoms of many elements are either permanently stable or else remain stable for intervals of time measured by millions of years, it would appear that these two conditions must be very approximately realized in

[1] Larmor: Aether and Matter, p. 233.

the constitution of many of the atoms of the different elements. Any atom in which these conditions were not fulfilled must long ago have disappeared and been broken up into more stable atomic systems.

It is thus not so much a matter of surprise that some atoms spontaneously break up, as that the atoms are such stable arrangements as they appear to be. The possibility of the disintegration of the atom is thus in most cases a necessary consequence of modern theories of atomic constitution.

Representations of Atomic Constitution

The recent developments in physical science have given a great impetus to the study of the constitution of the atom, and attempts have been made to form a mechanical, or rather electrical, representation of an atom which shall imitate as closely as possible the behavior of the actual atom.

On the electronic theory of matter it is supposed that the hydrogen atom consists of about a thousand electrons held in equilibrium by the internal forces of the atom. Since an atom is electrically neutral in regard to external bodies, it is necessary to assume that the negative charge carried by the electrons is compensated by the presence within the atom of an equal positive charge. The electrons are supposed to be the mobile parts of the atom, while the positive electricity is more or less fixed in position.

The earliest representation of such a model atom was given by Lord Kelvin.[1] A number of electrons or negatively charged particles were supposed to be arranged within a uniform sphere of positive electrification. The positive charge distributed throughout the sphere was equal in magnitude to the corresponding negative charges carried by the mobile electrons. This arrangement was very ingenious, for it not only fulfilled the condition that the atom was electrically neutral, but supplied the necessary forces within the atom to hold the electrons in equilibrium.

[1] Lord Kelvin: Phil. Mag., March, 1903; Oct., 1904; Dec., 1905.

Without such constraining forces it is obvious that the electrons would repel each other and escape from the atom. Lord Kelvin showed that certain arrangements of the electrons throughout the sphere were in stable equilibrium, while others were unstable, and a slight disturbance would lead either to their escape from the atom or to their falling into a more stable configuration. Lord Kelvin has recently devised certain arrangements of the positively and negatively charged particles constituting the atom which are unstable, and must lead to the expulsion of either a positively or negatively charged particle with great velocity, thus imitating the behavior of a radioatom in expelling α and β particles.

The conception of an atom, suggested by Lord Kelvin, was further developed by J. J. Thomson.[1] A number of electrons arranged in a ring at definite angular intervals were supposed to rotate uniformly within a sphere of positive electrification. He drew attention to a remarkable property of such a configuration. We have seen that a single electron moving in a circular orbit radiates energy, and the amount of the radiation becomes large when the electron is supposed to describe a circular orbit of atomic dimensions. When, however, a number of electrons followed one another in a circle, the fraction of the energy of motion of the electrons radiated per revolution decreased very rapidly as the number of electrons in the ring increased.

For example, the radiation from a group of six electrons moving with a velocity of one tenth that of light is less than one millionth of that of a single particle. For a velocity of one hundredth that of light the amount of radiation is only 10^{-16} of that from a single electron moving with the same velocity in the same orbit.

Such results show that an atom consisting of a number of rotating electrons may radiate energy extremely slowly, but ultimately this slow continuous drain of energy from the atom results in a diminution of velocity of the electrons. When this velocity falls below a certain critical value, the atom becomes

[1] J. J. Thomson: Phil. Mag., Dec., 1903; March, 1904.

unstable, and either breaks up with the expulsion of a part of the atom, or forms a new arrangement of the electrons.

J. J. Thomson considers that the cause of the disintegration of the atoms of radioactive matter must be ascribed to the loss of energy by radiation from the atom. He has mathematically investigated the possible temporarily stable arrangements of a given number of electrons within a sphere of uniform positive electrification. The properties of such an atom are very striking, and indirectly suggest a possible explanation of the periodic law of arrangement of the elements in chemistry. When the electrons revolve in one plane they tend to arrange themselves in a number of concentric rings, and generally, if free to move in any plane, in a number of concentric shells, like the coats of an onion.

It is not necessary here to consider in detail the arrangements discussed by J. J. Thomson, for these have already been given by him in the Silliman Lectures of two years ago. It suffices to say that such a model atom imitates in a remarkable way the behavior of the atoms of the elements, and also suggests a possible explanation of valency.

Some configurations of electrons, for example, are able to lose one, others two or more electrons, and yet remain stable. Others are able to gain an additional electron or two without altering the main features of the arrangement of the electrons. Those atoms which can readily lose electrons would correspond to an electropositive element, and *vice versa.*

Such attempts to imitate by an electrical model the structure of the atom are of necessity somewhat artificial, but they are of great value as indicating the general method of attack of the greatest problem that at present confronts the physicist. As our knowledge of atomic properties increases in accuracy it may yet be possible to deduce a structure of the atom which fulfils the conditions required by experiment. A promising beginning has already been made, and there is every hope that still further advances will soon be made in the elucidation of the mystery of atomic structure.

We have seen that on present theories positive electricity

plays a very different rôle from negative electricity. In order to hold the electrons together and to make the atom electrically neutral, it is necessary to call in the assistance of a fixed distribution of positive electrification. The mobile electrons constitute, so to speak, the bricks of the atomic structure, while the positive electricity acts as the necessary mortar to bind them together. This appears to be a somewhat arbitrary arrangement, but at present there appears to be no escape from this fundamental difficulty of the difference between positive and negative electricity.

Causes of Atomic Disintegration

We are now in a position to consider the possible causes that lead to the disintegration of the atoms of the radioelements. It has been shown that the law controlling the rate of disintegration of any individual product is very simple. The number of atoms breaking up per second is always in a constant ratio to the total number present. The value of this ratio, however, varies enormously for the different products. It has not been found possible to alter the rate of disintegration of any product by any external agency. Difference of temperature, which plays such an important part in altering the rate of chemical reactions, is entirely without influence on the rate of transformation of the radioactive bodies. For example, the heat emission of radium, which is a measure of the kinetic energy of the α particles, is unaltered by plunging the radium into liquid hydrogen. Elevation of temperature or chemical actions are equally without influence.

It thus appears that the atoms of the radioelements suffer spontaneous disintegration, or that it is brought about by forces beyond our control. It has been suggested that the atoms of radioactive matter may act as transformers of energy, abstracted in some way from the surrounding medium. Theories of this character were put forward for the special purpose of accounting for the emission of heat from radium without regard to the nature of the other radioactive processes. There is indubitable evidence that the heating effect of radium is a necessary consequence of the transformation of the radioatoms, and is a secondary effect resulting from the energy of motion of the expelled a particles.

Such theories do not take into account the fact that radioactivity is always accompanied by the appearance of new types of active matter. There must consequently be chemical changes in the active matter and, from other data, it is concluded that the changes occur in the atom itself and not in the molecule.

The causes which lead to the disintegration of the atom are at present a matter of conjecture. It is not yet possible to decide with certainty whether the disintegration is due to an external cause, or is an inherent property of the atom itself. It is conceivable, for example, that some unknown external force may supply the necessary disturbance to cause disintegration. In such a case, the external force supplies the place of a detonator to precipitate the atomic explosion. The energy liberated by the explosion, however, is derived mainly from the atom itself and not from the detonator. The law of transformation of radioactive matter does not throw any light on the question, for such a law is to be expected on either hypothesis.

It seems, however, most probable that the primary cause of atomic disintegration must be looked for in the atom itself, and consists in the loss of energy from the atom in the form of electromagnetic radiation. We have seen that unless certain conditions are fulfilled, an atom composed of negatively and positively charged particles will lose energy by radiation and ultimately break up.

For example, we have seen that J. J. Thomson has devised certain models of atoms which radiate energy extremely slowly, but which must ultimately, in consequence of the loss of atomic energy, become unstable and either break up or form a new atomic system. In the case of primary elements like uranium and thorium, the atoms are comparatively stable and have an average life of a thousand million years. The question then arises whether this radiation of energy is continuously occurring in all the atoms, or whether only a minute fraction is involved at one time. On the first view, all atoms formed at the same time should last for a definite interval. This, however, is contrary to the observed law of transformation, in which the atoms theoretically have a life embracing all values from zero to infinity.

We thus arrive at the conclusion that the configuration of the atom which gives rise to a radiation of energy only occurs in a minute fraction of the atoms present at one time, and is probably governed purely by the laws of probability.

There is one peculiarity of the transformation of the products of uranium, thorium, radium, and actinium which is possibly important in this connection. The β rays appear only in the last of the rapid series of changes of these elements, and these are expelled with enormous velocity. After the expulsion the resulting product is either permanently stable or far more stable than the preceding ones. It would appear more than a coincidence that the expulsion of a high velocity β particle should occur only in this final stage for each element. It is possible that the β particle, which is finally expelled, is the active agent in promoting the previous transformations, and that when once the disturbing factor has been removed, the resulting atom sinks into a configuration of far more stable equilibrium.

For example, one of the electrons composing the atom may take up a position in the atomic system which leads to a radiation of energy. As a result the atom breaks up with the expulsion of an α particle, and this process continues through successive stages until, finally, there is a violent explosion within the atom, which results in the expulson of the disturbing electron with enormous speed.

Processes occurring in Radium

Consider a minute quantity of radium of weight about one millionth of a milligram in radioactive equilibrium. This will contain about 3.6×10^{12} atoms of radium of atomic weight 225. Since in one gram of radium 6.2×10^{10} α particles are expelled per second from the radium itself, the number disintegrating per second in one millionth of a milligram is 62. On an average, an exactly equal number of α particles will be expelled from each of the successive α ray products, viz., the emanation, radium A, and radium C.

The number of atoms of each of the radioactive products present with this quantity of radium will be very different. For 3.6×10^{12} atoms of radium, there will be about 3×10^{7} of atoms of emanation, 1.6×10^{4} atoms of radium A, 1.5×10^{5} of radium B, and 1.15×10^{5} of radium C. The radium atoms present will thus enormously preponderate over those of its products.

Supposing it were possible to magnify this small particle of radium so as to distinguish the individual atoms, we should see a large number of radium atoms, and mixed with them a very small number of atoms of its products; but if the attention were focussed on the atoms of each of the individual substances present we should find that the same number of α particles are expelled from each of them per second. The number of atoms of each product remains on an average constant, for the supply of fresh atoms compensates for those that break up.

The electronic theory of matter supposes that an atom is composed of a swarm of electrons in rapid movement held in equilibrium by the internal forces of the atom. In the case of heavy atoms, like those of the radioelements, it is not necessary to suppose that each of these electrons has complete freedom of movement. The character of the transformation of the atom suggests that it is built up in part of a number of secondary units, consisting of groups or aggregates of electrons in equilibrium, which are in rapid independent motion within the atom.

For example, it seems probable that the α particle or helium atom actually exists as an independent unit of matter within the radium atom, and is released at the moment of the disintegration of the latter. These α particles are in rapid movement, and when a stage of instability is reached, one is ejected from the atom with the velocity it possessed in its atomic orbit. If this be the case, the α particles, on an average, must have a velocity within the atom of more than 1/30 that of light.

It is possible, however, that a part of their great kinetic energy may be acquired during the process of expulsion from the atom, for we have seen that, from a variety of considerations, the atom must be supposed to be the seat of intense electrical forces. At the moment preceding the expulsion of an α particle, from radium for example, the atom must be in a state of violent disturbance. As a result, the forces which constrain one of these α particles within the atom are momentarily neutralized and the α particle escapes from the atom at an enormous speed.

The internal forces are still sufficiently powerful to prevent the escape of the other parts of the atom, and there is a rapid adjustment of its components to form a new system which is temporarily stable. It is probable that for a short interval after the escape of the α particle, the atom is in a state of violent disturbance, but finally sinks again into a temporarily stable system. The residual atom has smaller mass than before, and the internal arrangement of its parts is entirely different from the previous one. The new atom, in fact, becomes an atom of the emanation, and has chemical and physical properties entirely different from those of the parent atom.

The atoms of the emanation are not nearly so stable as those of radium, for they have an average life of only six days. The atom breaks up as before, expelling another α particle and giving rise to an atom of radium A, which again differs widely in properties from those of the emanation and of radium. This substance is very unstable, for the atoms have an average life of only four minutes. After the loss of another α particle the atom of radium B makes its appearance. This atom undergoes a change, which is apparently different in character from the others. It may or may not expel an α particle, but if it does so, the particle travels at too low a speed to produce appreciable ionization in the gas. The atom, however, expels a β particle at a moderate speed. The atom of radium B then changes into C. The instability of the latter atom results in an explosion of great intensity. An α particle is expelled at a greater speed than from any other product, while at the same time a β particle is ejected with a velocity nearly equal to that of light. After this violent explosion the resulting atom sinks into a far more stable system, but this eventually breaks up, as we have seen, and passes through still further stages. Finally, after the expulsion of an α particle from radium F, the resulting atom is probably identical with that of lead.

Such considerations show that a mass of radium is the seat of an extraordinary conflict of forces. In a gram of radium, for example, about 6×10^{10} α particles are shot out each second from each of the α ray products, while in addition there is an expulsion of an equal number of high velocity electrons from radium B and C. Since the α particles are only able to pass through a small thickness of matter, the greater number of these α particles are stopped in the radium itself, which is consequently exposed to a bombardment of great intensity.

Let us concentrate our attention for a moment on an atom of radium at the moment of the expulsion of an α particle. If the principle of equivalence of momenta holds, the expulsion of the α particle must cause a recoil of the atom from which it escapes. Since the α particle has a mass of about 1/50 that of the radium atom, and is expelled with a velocity of nearly 2×10^9 cms. per second, the atom of radium must recoil with an initial velocity of about 4×10^7 cms. per second, or about 200 miles per second. This velocity will decrease rapidly in consequence of the collisions of the moving atom with the atoms in its path, and it will probably be brought nearly to rest after traversing a very small distance. The kinetic energy of motion of the radium atom will thus be transformed into heat. The α particle initially starts with an enormous velocity, and must force its way through the atoms of radium in its path, knocking off from them a shower of electrons in the process. Its energy is gradually used up in producing ions, and its velocity consequently diminishes. Finally, it loses its power of ionizing, and is brought to rest. Its charge is neutralized, and the α particle then becomes a helium atom, and is mechanically imprisoned in the mass of the radium. The energy used up in producing ions in the radium is finally given up in the form of heat, for the ions recombine, emitting heat during the process.

Let us now turn our attention to the processes occurring in

the air or other gas surrounding the radium. The *a* particles expelled from the surface of a mass of radium escape into the air without loss of speed due to traversing the radium, and the *a* particle from each product has its characteristic velocity. Imagine that we can follow visually the flight of an *a* particle through the gas. The velocity of the *a* particle is initially so great compared with the velocity of translation of the molecules of the gas, that the latter will appear to stand still during the flight of the *a* particle. The time is too short for the molecules of air to escape from the path of the *a* projectile, and its velocity and energy are so great that it is able to plunge through the molecules in its path. The electric disturbance produced by its passage through the molecule may lead to the expulsion of an electron, and, probably in many cases, to the breaking up of the complex molecule into charged atoms.

Two or more ions are consequently produced as a result of the passage through the molecule. This process continues until, after passing through about 3.5 cms. of air under normal conditions, and producing about 100,000 ions, its power of ionization is lost. Exactly what happens to the *a* particle at the end of its career of ionization is not yet known. As we have seen, experiment indicates that some of the *a* particles are still moving at a high velocity when their ionizing power becomes very small. Since the initial rapid reduction of the velocity of the *a* particle appears to be mainly a result of the energy used up in ionizing the gas, it is probable that the *a* particle, after its power of ionizing has been largely lost, will traverse a considerable distance of air before it is brought to rest by continued collision with the gas molecules. We are unable to detect the presence of such *a* particles, since they have lost all the properties which serve ordinarily to detect their presence.

There is a considerable amount of evidence to show that the energy absorbed from the *a* particle in producing a pair of ions is much greater than that required merely to separate the positive ion from the negative. Such a result suggests that during the process of ionization the ions acquire a considerable velocity, and that the energy spent in setting the ions in motion is large compared with that necessary for the mere separation of the ions from the immediate sphere of each other's influence.

Although the β particle escapes from the radium with an average velocity ten times that of the α particle, it is far less efficient as an ionizer. It produces only a small number of ions per centimeter of path compared with the number produced by an α particle, and passes through about 100 times the distance in air before it ceases to ionize.

We have not so far considered the connection of the γ rays with radioactive changes. These rays always appear with the β rays and are believed to be an electromagnetic pulse, set up in consequence of the sudden expulsion of the β particle. This pulse is the seat of very intense electric and magnetic forces and travels out from the atom like a spherical wave with the velocity of light. Such a pulse is a very inefficient ionizer compared with the α particle, and on an average produces only one ion per centimeter of its path for 10,000 produced by the α particle. The penetrating power of the γ ray, on the other hand, is very great, and it continues to produce ions even after traversing a great distance of air.

The energy of the rays which traverse the gas is ultimately frittered down into heat. The initial energy of motion of the ions is rapidly lost by collision with the gas particles, while the ions finally recombine with the liberation of energy.

In addition to the ionization effects already considered, there are very marked secondary effects produced when the rays impinge upon matter. The α particles release a shower of electrons from the matter upon which they impinge. These electrons, however, are emitted at a very slow speed. On the other hand, the β and γ rays cause the release of electrons at a speed comparable with that of light. These secondary radiations are most marked when the radiations fall on heavy metals like lead, but no doubt occur, though in a much less intense degree, during the passage of the rays through a gas.

On account of the great energy of motion of the α particle, it might be expected to set in vibration the atoms of matter in its path, and cause them to emit light waves. This property of the *a* rays of exciting luminosity was first noted by Sir William and Lady Huggins,[1] who found that the weak phosphorescent light of radium showed the band spectrum of nitrogen. This has been traced to the action of the *a* rays, either in free nitrogen close to the radium, or in nitrogen occluded within the radium compound. Such a result is of unusual interest, as it is the first example of a gas giving a spectrum when cold without the stimulus of a strong electric discharge.

Walter and Pohl[2] have recently found that the gas traversed by the rays from an active preparation of radiotellurium emits light waves which act on a photographic plate. The intensity of this effect is greatest for pure nitrogen. The atoms of nitrogen appear to be more easily stimulated to give out their characteristic vibrations than any other gas so far examined. It is a matter of surprise that as yet no evidence has been obtained that the *a* particles themselves give a spectrum. The violent collisions of the *a* particle with the molecules in its path must set up vibrations in the *a* particle, and it should give a characteristic spectrum. Experiments of this character, though of great difficulty, are most important, for they may throw light on the nature of the *a* particle. In this connection it is of interest to note that Giesel found that a preparation of "emanium" emitted a phosphorescent light consisting of bright lines. These lines were found to be due to didymium, which was present as an impurity in the active substance.

There is no doubt that the radiations emitted from active bodies serve as very powerful agents for the ionization and dissociation of matter. No definite evidence has so far been obtained that the *a* or *β* particles emitted from radium are able to hasten its rate of transformation. Such a concentrated source of energy as these high velocity particles might be expected to produce, under some conditions, a disintegration of the atoms of matter through which they pass. A mass of radium, for example, which is subjected to an intense bombardment by its own *a*

[1] Sir William and Lady Huggins: Proc. Roy. Soc., lxxii, pp. 196, 409 (1903); lxxvii, p. 130 (1906).

[2] Walter and Pohl: Ann. d. Phys., xviii, p. 406 (1905).

and β particles might be expected to disintegrate faster than the same amount of radium diffused through a large volume. Further experiments in this direction may yet show that such an effect does exist, but it is certainly not very marked. A direct attack on the question as to whether X-rays are able to cause the disintegration of matter has been made by Bumstead.[1] A strong beam of X-rays fell on two plates of zinc and lead, which were of such a thickness as to absorb an equal fraction of energy of the incident beam. The lead was raised to a considerably higher temperature than the zinc, indicating that although the same fraction of the energy of the rays had been absorbed, more energy had been released in the lead than in the zinc. Such a result suggests that the X-rays cause a greater amount of atomic disintegration in lead than in zinc, and that a considerable portion of the heat generated in the lead is due to the energy liberated from the transformation of its atoms.

Further experiments with a variety of metals and with different sources of intense ionizing radiations are required to substantiate completely such a far-reaching conclusion, but the results so far obtained in this difficult field of research certainly lead us to hope that we may yet bring about the disintegration of atoms by laboratory methods.

We have previously indicated that there is a very strong proof that ordinary matter possesses the property of emitting characteristic radiations which are able to ionize a gas. Such results suggest that there is an extremely slow transformation of matter of a type similar to that shown by the radioactive bodies. It is not necessary to suppose that the α particles from all types of matter is the same mass. For example, hydrogen may be expelled from some bodies instead of helium. The experimental observation that the α particle loses its power of acting on a photographic plate and of ionizing the gas when its velocity falls to about 8×10^8 cms. per second is of great importance in this connection. There is no doubt that if α particles were expelled from matter below this velocity they would produce very little if any electrical effect. It is certainly a matter of remark

[1] Bumstead: Phil. Mag., Feb., 1906.

that the average *a* particle from radioactive bodies is projected at a speed less than twice this minimum velocity. It appears by no means improbable that the so-called radioactive bodies may differ from ordinary matter mainly in their power of expelling *a* particles above this critical speed. Ordinary matter which produces extremely weak ionizing effects might be emitting *a* particles at a rate comparable with uranium, but yet, if their power of expulsion was less than this critical value, it would be difficult to detect their presence.

Such considerations show that it is by no means necessary to suppose that the transformation of matter should always be accompanied by the intense electrical and other effects exhibited by the radioactive bodies proper. Matter may be undergoing slow atomic transformation of a character similar to radium, which would be difficult to detect by our present methods.

INDEX.

INDEX